Nuclear Threats and Security Challenges

NATO Science for Peace and Security Series

This Series presents the results of scientific meetings supported under the NATO Programme: Science for Peace and Security (SPS).

The NATO SPS Programme supports meetings in the following Key Priority areas: (1) Defence Against Terrorism; (2) Countering other Threats to Security and (3) NATO, Partner and Mediterranean Dialogue Country Priorities. The types of meeting supported are generally "Advanced Study Institutes" and "Advanced Research Workshops". The NATO SPS Series collects together the results of these meetings. The meetings are co-organized by scientists from NATO countries and scientists from NATO's "Partner" or "Mediterranean Dialogue" countries. The observations and recommendations made at the meetings, as well as the contents of the volumes in the Series, reflect those of participants and contributors only; they should not necessarily be regarded as reflecting NATO views or policy.

Advanced Study Institutes (ASI) are high-level tutorial courses to convey the latest developments in a subject to an advanced-level audience

Advanced Research Workshops (ARW) are expert meetings where an intense but informal exchange of views at the frontiers of a subject aims at identifying directions for future action

Following a transformation of the programme in 2006 the Series has been re-named and re-organised. Recent volumes on topics not related to security, which result from meetings supported under the programme earlier, may be found in the NATO Science Series.

The Series is published by IOS Press, Amsterdam, and Springer, Dordrecht, in conjunction with the NATO Emerging Security Challenges Division.

Sub-Series

A.	Chemistry and Biology	Springer
B.	Physics and Biophysics	Springer
C.	Environmental Security	Springer
D.	Information and Communication Security	IOS Press
E.	Human and Societal Dynamics	IOS Press

http://www.nato.int/science
http://www.springer.com
http://www.iospress.nl

Series B: Physics and Biophysics

Nuclear Threats and Security Challenges

edited by

Samuel Apikyan
National Center of Security and Protection, USA

and

David Diamond
Brookhaven National Laboratory, USA

Springer
Published in Cooperation with NATO Emerging Security Challenges Division

Proceedings of the NATO Advanced Research Workshop on
Preparedness for Nuclear and Radiological Threats
Los Angeles, USA
18–20 November 2014

Library of Congress Control Number: 2015944112

ISBN 978-94-017-9936-2 (PB)
ISBN 978-94-017-9893-8 (HB)
ISBN 978-94-017-9894-5 (e-book)
DOI 10.1007/978-94-017-9894-5

Published by Springer,
P.O. Box 17, 3300 AA Dordrecht, The Netherlands.

www.springer.com

Printed on acid-free paper

Preface

The world faces no greater or more urgent danger than a terrorist attack with the intent of killing, maiming, and traumatizing a large population. International peace and security is threatened in particular by the proliferation of nuclear materials and technologies that could lead to a nuclear or radiological attack. More nations are trying to acquire nuclear weapons, and black markets trade in nuclear secrets and materials. Terrorists are determined to buy, build, or steal a nuclear weapon or use a radioactive source in a conventional bomb.

Organizations like al Qaeda and the so-called Islamic State have said that obtaining these weapons and perpetrating another "Hiroshima" are their "religious duty." Organizations such as these have the will, the technical know-how, and the financial resources to make these threats a reality.

Our strategy to combat these threats is multilayered, and events in recent years have shown the necessity to continually reevaluate national preparedness programs. Throughout the world there are people working on the key issues related to this subject such as:

- Preventing, avoiding, or stopping threats
- Protecting our citizens and assets against the greatest threats and hazards
- Mitigating the loss of life and property by lessening the impact of future disasters
- Responding quickly to save lives, protect property and the environment, and meet basic human needs in the aftermath of a catastrophic incident
- Recovering through timely restoration and strengthening of infrastructure and the economy, as well as the social fabric of communities affected by a catastrophic incident

The NATO Advanced Research Workshop on "Preparedness for Nuclear and Radiological Threats" was held in Los Angeles, on 18–20 November 2014 with support from the NATO Science for Peace and Security Programme. The purpose of the workshop was to contribute to the critical assessment of existing knowledge on this subject, to identify directions for future research and policies, and to promote close working relationships between scientists, engineers, and policy makers from different countries and with different professional experience. More than 100 representatives of 18 countries participated. The program was built upon the accomplishments of The Hague 2014 Nuclear Security Summit and previous NATO workshops such as "Countering Nuclear/Radiological Terrorism" (2005); "Prevention, Detection and Response to Nuclear and Radiological Threat" (2007); and "Threat Detection, Response and Consequence Management Associated with Nuclear and Radiological Terrorism" (2008).

This book contains approximately half of the papers presented at the workshop. The other half of the papers are found in the book *Nuclear Terrorism and National Preparedness*. We hope it will be useful not only for the multinational scientific and technical communities engaged in combating nuclear and radiological terrorism but also for decision makers and for those working at governmental and policy levels whose actions affect the directions the science takes and how the technology is incorporated into country-specific national systems for combating nuclear and radiological threats.

Los Angeles — Samuel Apikyan
Upton — David Diamond

Contents

Part I
Nuclear Security and Nonproliferation

Chapter 1
GAO: Two Decades Evaluating the Impact and Effectiveness of U.S. Nuclear and Radiological Material Security Programs

David Trimble

Abstract Since the early 1990s, the Government Accountability Office (GAO), the investigative arm of the U.S. Congress, has been reporting on the impact and efficiency of numerous federal programs—that have collectively cost U.S. taxpayers billions of dollars—to reduce the risks posed by vulnerable nuclear and radiological materials worldwide. GAO's assessments have focused on, among other things, the Department of Energy's (DOE) Material Protection, Control and Accounting program in Russia and the Global Threat Reduction Initiative that has been implemented in more than 100 countries. More recently, GAO has assessed federal agencies' efforts and strategies to implement the President's initiative to secure all vulnerable material worldwide within a 4-year period. A significant and growing part of GAO's portfolio, particularly after September 11, focuses on radiological material security, including federal preparedness for and response to a terrorist attack involving either a radiological dispersal device or improvised nuclear device attack in the United States.

An independent, nonpartisan agency, GAO's mission is to support the Congress in meeting its constitutional responsibilities and to help improve the performance and ensure the accountability of the federal government for the benefit of the American people. GAO provides the Congress with timely information that is objective, fact-based, nonpartisan, nonideological, fair, and balanced. Within GAO, the U.S. and International Nuclear Security and Cleanup mission group covers a wide range of nuclear issues that include nuclear nonproliferation; nuclear and radiological smuggling and terrorism; and special nuclear material production, consolidation, and storage. This paper provides an overview of the recent nuclear nonproliferation work that GAO has undertaken on behalf of the Congress, focusing on highlights of key reports, major findings, recommendations, and impact on federal agencies' programs.

D. Trimble (✉)
Government Accountability Office, Washington, DC, USA
e-mail: trimbled@gao.gov

S. Apikyan, D. Diamond (eds.), *Nuclear Threats and Security Challenges*, NATO Science for Peace and Security Series B: Physics and Biophysics, DOI 10.1007/978-94-017-9894-5_1

1.1 Introduction

One of the most serious threats facing the United States and other countries is the possibility that other nations or terrorist organizations could steal a nuclear warhead or nuclear weapon-usable materials from poorly secured stockpiles around the world, or that nations could divert nuclear material intended for peaceful purposes to the development of nuclear weapons.[1] Of great concern is that terrorists could fashion a crude nuclear bomb made from either highly enriched uranium (HEU) or plutonium into an improvised nuclear device. Such a device would create an explosion producing extreme heat, powerful shockwaves and intense radiation that would be immediately lethal to individuals within miles of the explosion, as well as radioactive fallout over thousands of square miles. Nonproliferation experts estimate that a successful improvised nuclear device could devastate the heart of a medium-sized U.S. city and could cause hundreds of thousands of deaths and injuries, as well as pose long-term cancer risks to those exposed to the radioactive fallout.

Radiological material also poses a significant security threat to the United States and the international community. Radiological material—such as cobalt-60, cesium-137, and iridium-192—is encapsulated or sealed in metal to prevent its dispersal and is commonly called a sealed radiological source. Sealed radiological sources are used worldwide for many legitimate purposes, such as medical, industrial, and agricultural applications. The total number of these sources in use worldwide is unknown because many countries do not systematically account for them. If certain types of these sources were obtained by terrorists, they could be used to produce a simple and crude but potentially dangerous weapon—known as a radiological dispersion device, or dirty bomb. Although experts believe that a dirty bomb would result in a limited number of deaths, it could have severe economic consequences. Depending on the type, amount, and form, the dispersed radiological material could cause radiation sickness for people nearby and produce serious economic and psychological disruption associated with the evacuation and subsequent cleanup of the contaminated area.

1.2 The Post-Cold War Era

GAO has been reporting on these and related nonproliferation and nuclear security issues since the 1990s. In the years following the end of the Cold War, GAO's work in this area focused on, among other things, U.S. nuclear engagement with the former Soviet Union. For example, reports from that time addressed the status of U.S. efforts to improve nuclear material controls in the newly independent

[1] Weapon-usable nuclear materials are highly enriched uranium, uranium-233, and any plutonium containing less than 80 % of the isotope plutonium-238. Such materials are also often referred to as fissile materials or strategic special nuclear materials.

states of the former Soviet Union, the status of transparency measures for the U.S. purchase of Russian highly enriched uranium, and DOE's efforts to mitigate the risk to nonproliferation goals of unemployed former Soviet Union weapons scientists.[2] GAO also published several reports during this time on the International Atomic Energy Agency's (IAEA) nonproliferation activities, including reviews of the uncertainties regarding IAEA's changing safeguards system and the agency's ability to monitor operations at North Korean nuclear facilities.[3] Below are two examples of this work.

- *Status of U.S. Efforts to Improve Nuclear Material Controls in Newly Independent States*: In 1996, GAO reviewed U.S. efforts to strengthen controls over nuclear materials in the newly independent states of the former Soviet Union.[4] The Soviet Union produced about 1,200 metric tons of highly enriched uranium and 200 metric tons of plutonium, with much of this material outside of nuclear weapons and highly attractive to theft. GAO found that, at the time, the newly independent states may not have had accurate and complete inventories of the material they inherited and that the breakdown of Soviet-era material protection, control, and accounting (MPC&A) systems may have left the newly independent states unable to counter the threat of theft. In addition, GAO found that nuclear facilities could not quickly detect and localize nuclear material losses or detect unauthorized attempts to remove nuclear material, and seizures of direct-use material in Russia and Europe had increased concerns about theft and diversion. GAO also found that U.S. agencies had begun efforts to help the newly independent states improve their MPC&A systems for direct-use material, as well as cooperation with Russia's nuclear regulatory agency to develop a national MPC&A regulatory infrastructure.
- *Uncertainties with Implementing IAEA's Strengthened Safeguards System*: In 1998, GAO reviewed changes IAEA was undertaking at the time to strengthen its safeguards program by introducing advanced safeguards techniques under its existing safeguards agreements.[5] IAEA has a dual role of promoting the peaceful uses of nuclear energy through its nuclear safety and technical cooperation programs, and verifying, through its safeguards program, that nuclear

---

[2]For example, see *Nuclear Nonproliferation: Status of U.S. Efforts to Improve Nuclear Material Controls in Newly Independent States*, GAO/NSIAD/RCED-96-89 (Washington, D.C.: Mar. 8, 1996); *Nuclear Nonproliferation: Status of Transparency Measures for U.S. Purchase of Russian Highly Enriched Uranium*, GAO/RCED-99-194 (Washington, D.C.: Sep. 22, 1999); and *Nuclear Nonproliferation: Concerns with DOE's Efforts to Reduce the Risks Posed by Russia's Unemployed Weapons Scientists*, GAO/RCED-99-54 (Washington, D.C.: Feb. 19, 1999).

[3]For example, see *Nuclear Nonproliferation: Uncertainties with Implementing IAEA's Strengthened Safeguards System*, GAO/NSIAD/RCED-98-184 (Washington, D.C.: Feb. 10, 1999) and *Nuclear Nonproliferation: Difficulties in Accomplishing IAEA's Activities in North Korea*, GAO/RCED-98-210 (Washington, D.C.: Jul. 7, 1998).

[4]GAO/NSIAD/RCED-96-89.

[5]GAO/NSIAD/RCED-98-184.

materials subject to safeguards are not diverted to nuclear weapons or other proscribed purposes. In response to Iraq's secret nuclear weapons program, the international community, led by the United States, launched an intensive effort to create a new capability within the IAEA's safeguards system to detect secret or undeclared activities. IAEA also sought additional rights to conduct more intrusive inspections and collect information on nuclear activities through an Additional Protocol that supplemented the existing safeguards agreements. These changes to the agency's safeguards systems were intended to give its inspectors greater ability to detect clandestine nuclear activities in non-nuclear weapons states that are signatories to the Non-Proliferation of Nuclear Weapons or other regional nonproliferation treaties. GAO reported that, under existing safeguards agreements with states and regional organizations, IAEA had increased its access to information on all nuclear activities at declared facilities in non-nuclear weapons states. However, we recommended that IAEA develop and circulate a plan for implementing elements of the enhanced safeguards system. GAO most recently reported on the status of IAEA's safeguards program in 2013.[6]

1.3 The Post-September 11 Era

Following the terrorist attacks of September 11, 2001, U.S. and international experts raised concerns that unsecured radiological sources were vulnerable to theft and posed a significant security threat to the United States and the international community. In 2003, GAO issued a number of reports focusing on U.S. and international efforts to secure radiological sources and recover unwanted sources.[7] In 2007, GAO issued a report showing that many of the highest-risk and most dangerous sources still remain unsecured, particularly in Russia.[8]

- *U.S. and International Assistance Efforts to Control Sealed Radiological Sources*: In 2003, GAO reviewed the number of sealed sources in use worldwide, as well as those that have been lost, stolen, or abandoned.[9] GAO found that the

[6]*Nuclear Nonproliferation: IAEA Has Made Progress in Implementing Critical Programs but Continues to Face Challenges*, GAO-13-139 (Washington, D.C.: May 16, 2013).

[7]For example, see *Nuclear Nonproliferation: DOE Action Needed to Ensure Continued Recovery of Unwanted Sealed Radioactive Sources*, GAO-03-483 (Washington, D.C., April 15, 2003), *Nuclear Nonproliferation: U.S. and International Efforts to Control Sealed Radioactive Sources Need Strengthening*, GAO-03-638 (Washington, D.C., May 16, 2003), and *Nuclear Security: Federal and State Action Needed to Improve Security of Sealed Radioactive Sources*, GAO-03-804 (Washington, D.C., Aug. 6, 2003).

[8]For example, see *Nuclear Nonproliferation: DOE's International Radiological Threat Reduction Program Needs to Focus Future Efforts on Securing the Highest Priority Radiological Sources*, GAO-07-282 (Washington, D.C., Jan. 31, 2007).

[9]GAO-03-804.

precise number of sealed sources is unknown because many countries do not systematically account for them. However, at the time, nearly ten million sealed sources existed in the United States and the 49 countries responding to a GAO survey. There is also limited information about the number of sealed sources that have been lost, stolen, or abandoned, but it was estimated to be in the thousands worldwide. Furthermore, many of the most vulnerable sealed sources that could pose a security risk were located in the countries of the former Soviet Union. GAO recommended in the report, among other things, that the Secretary of Energy take the lead in developing a comprehensive plan to strengthen controls over other countries' sealed sources.

- *DOE's International Radiological Threat Reduction Program*: In 2007, GAO assessed the progress DOE had made in implementing its program to help other countries secure their sealed radiological sources, as well as described DOE's coordination with other U.S. agencies and international organizations to secure radiological sources in other countries.[10] GAO found that, since 2002, DOE had upgraded the security of hundreds of sites in other countries that contained radiological sources and had achieved noteworthy accomplishments, including removing radioactive material in Chechnya. However, DOE had made limited progress in securing many of the most dangerous sources located in waste storage facilities and hundreds of sources across Russia contained in radioisotope thermoelectric generators (RTG). As a result, as of September 2006, almost 70 % of all sites secured were medical facilities, which generally contain one radiological source, and many of the highest-risk and most dangerous sources still remained unsecured, particularly in Russia. For example, GAO reported that 16 of 20 waste storage sites across Russia and Ukraine remained unsecured, while more than 700 RTGs remained operational or abandoned in Russia and were vulnerable to theft or potential misuse. In the report, GAO made several recommendations to DOE to better prioritize sites to be selected for security upgrades and strengthen program management practices.

1.4 Recent GAO Work

In recent years, GAO's nonproliferation work has continued to focus on the security of vulnerable nuclear materials worldwide. Recent reports have included a preliminary assessment of the President's 4-year global nuclear security initiative and U.S. agencies' ability to account for U.S. nuclear material overseas. GAO has also increased its focus on the security of radiological materials and has produced recent reports and testimonies on the security of radiological sources in hospitals and in industrial use in the United States. The following are summaries of some recent key reports.

---

[10]GAO-07-282.

- *Government-wide Strategy to Implement the President's 4-Year Global Nuclear Material Security Initiative Lacked Important Details*: In December 2010, GAO reported on aspects of U.S. planning and strategies to secure all vulnerable nuclear materials worldwide within a 4-year period in response to a new international initiative announced by the President in 2009.[11] Despite individual agency efforts to implement the 4-year initiative, GAO found that the overarching interagency strategy—which the National Security Council (NSC) coordinated—lacked specific details concerning how the initiative would be implemented, including the identity of, and details regarding, vulnerable foreign nuclear material sites and facilities to be addressed; potential challenges and strategies for overcoming these challenges; and cost estimates. GAO found that, absent such an implementation plan, essential details associated with the 4-year initiative were unclear, including the initiative's overall estimated costs, time frames, and scope of work. GAO recommended that NSC lead and coordinate the development of a comprehensive plan for implementing this initiative—one that, among other things, clearly identified the specific foreign countries, sites, and facilities where materials were determined to be poorly secured; planned activities, potential implementation challenges, and steps needed to overcome those challenges at each location; and estimated time frames and costs associated with achieving the 4-year goal. NSC did not comment on GAO's recommendation, which has still not been implemented as of October 2014.
- *U.S. Agencies Have Limited Ability to Account for, Monitor, and Evaluate the Security of U.S. Nuclear Material Overseas*: The United States has exported special nuclear material, including enriched uranium, and source material such as natural uranium under nuclear cooperation framework agreements for many years.[12] In September 2011, GAO issued a report that assessed U.S. agency efforts to account for U.S. nuclear material overseas and efforts to evaluate the security of these materials.[13] In that report, GAO found that U.S. agencies—DOE, the Nuclear Regulatory Commission (NRC), and the Department of State (State)—are not able to fully account for U.S. nuclear material overseas that is subject to the terms of nuclear cooperation agreements because the agreements do not stipulate systematic reporting of such information, and there is no policy to pursue or obtain such information. These agreements generally require that partners report inventory information upon request. However, GAO reported that the agencies had not systematically sought such information, and suggested

[11]GAO, *Nuclear Nonproliferation: Comprehensive U.S. Planning and Better Foreign Cooperation Needed to Secure Vulnerable Nuclear Materials Worldwide*, GAO-11-227 (Washington, D.C.: Dec. 15, 2010).

[12]The United States has 24 nuclear cooperation agreements in force for peaceful civilian cooperation with partners, including foreign countries, the European Atomic Energy Community (EURATOM), IAEA, and Taiwan.

[13]GAO, *Nuclear Nonproliferation: U.S. Agencies Have Limited Ability to Account for, Monitor, and Evaluate the Security of U.S. Nuclear Material Overseas*, GAO-11-920 (Washington, D.C.: Sept. 8, 2011).

that the Congress consider directing DOE and NRC to compile an inventory of U.S. weapon-usable nuclear materials overseas. GAO also found that DOE has taken steps to improve security at a number of facilities overseas that hold U.S. nuclear material but faces constraints. For example, DOE removes U.S. material from vulnerable facilities but can only repatriate materials that have an approved disposition pathway and meet the program's eligibility criteria. Officials told us that of the approximately 17,500 kg of HEU exported from the United States, 12,400 kg were not eligible for return to the United States, and the majority of that was not eligible because the material does not have an acceptable disposition pathway. GAO recommended, among other things, that State. DOE, and NRC work together to determine baseline inventories of U.S. weapons-usable nuclear materials for partners that do not already have an annual inventory reconciliation. GAO also recommended that those agencies establish reporting on a facility-by-facility basis for those partners that already have an annual inventory reconciliation. To date, agencies have taken some actions on these recommendations.

- *Agencies Have Taken Steps to Secure Domestic Radiological Materials, but Gaps Remain*: Radiological material is commonly found in equipment used by U.S. medical facilities to treat, among other things, cancer patients. In 2008, DOE's National Nuclear Security Administration (NNSA) established a program to provide security upgrades to U.S. hospitals and medical facilities that use radiological sources. In September 2012, GAO reported on issues related to the extent to which NRC's requirements ensure the security of radiological sources at U.S. medical facilities and the status of NNSA's efforts to improve the security of sources at these facilities.[14] GAO found that NRC's requirements did not consistently ensure the security of high-risk radiological sources at the 26 selected hospitals and medical facilities GAO visited for that review. One reason for this is that the requirements are broadly written and do not prescribe specific measures that hospitals and medical facilities must take to secure medical equipment containing sealed sources. Rather, the requirements provide a general framework for what constitutes adequate security practices, which is implemented in various ways at different hospitals. While NNSA has conducted outreach efforts in partnership with NRC and Agreement States to encourage participation in its security upgrade program, there are still many facilities that are not participating. Regarding NNSA's security upgrades, of the 26 hospitals and medical facilities that GAO visited, 13 had volunteered for and received security upgrades, such as remote monitoring systems, surveillance cameras, and tamper alarms; three others were in the process of receiving upgrades. However, the report noted that the program's impact is limited because, among other things, it is voluntary, and facilities can decline to participate. GAO recommended, among other things, that NRC strengthen its security requirements by providing medical facilities with

[14]GAO, *Nuclear Nonproliferation: Additional Actions Needed to Improve Security of Radiological Sources at U.S. Medical Facilities*, GAO-12-925 (Washington, D.C.: Sep. 10, 2012).

specific measures they must take to develop and sustain a more effective security program. NRC neither agreed nor disagreed with this recommendation and stated that its existing security requirements are adequate. GAO continues to believe that implementing this recommendation would contribute to increased security at U.S. hospitals and medical facilities.

- *Additional Actions Needed to Increase the Security of U.S. Industrial Radiological Sources*: Following GAO's work on security of radiological sources in hospitals, GAO issued a report in June 2014 on potential security risks with radiological sources in the industrial sector.[15] GAO found that challenges exist in reducing the security risks faced by licensees using high-risk industrial radiological sources, including at industrial facilities in the oil and gas, aerospace, and food sterilization sectors.[16] The challenges licensees face include the portability of some radiological sources (i.e., mobile sources), which makes them susceptible to theft or loss, as the size of some of these sources is small enough for them to be easily concealed. The most common mobile source is contained in a device called a radiography camera. GAO identified four incidents from 2006 to 2012 where such cameras were stolen. Three of the four stolen sources were recovered, and one of the stolen sources was never recovered. These thefts occurred after NRC has established increased security controls in 2005. Licensees also face challenges in protecting against an insider threat—determining which employees are suitable for trustworthiness and reliability (T&R) certification to have unescorted access to high-risk radiological sources. GAO found two cases where employees were granted unescorted access despite having serious criminal records. In one of the cases, the individual had been twice convicted of terroristic threats. NRC officials said that the person was convicted not of a threat against the United States, but of making violent verbal threats against two individuals. It is unclear whether these cases represent isolated incidents or a systemic weakness in the T&R process established by NRC. The report found that federal agencies responsible for securing radiological sources were taking steps to better secure industrial radiological sources. For example, NRC, which is responsible for licensing and regulating the commercial use of radiological sources, has been developing a best practices guide. In addition, NNSA, which provides voluntary security upgrades to facilities with such sources, had two initiatives under way to improve industrial radiological source security. The report found, however, that NRC, NNSA, and DHS—agencies that play a role in nuclear and radiological security—were not always effectively collaborating to achieve the common mission of securing industrial sources.

[15]GAO, *Nuclear Nonproliferation: Additional Actions Needed to Improve Security of U.S. Industrial Radiological Sources*, GAO-14-293 (Washington, D.C.: June 6, 2012).

[16]A licensee is a company, organization, institution, or other entity to which NRC or state agencies have granted a general license or specific license to construct or operate a nuclear facility, or to receive, possess, use, transfer, or dispose of source material, by-product material, or special nuclear material.

GAO recommended that agencies review their existing efforts for opportunities to enhance collaboration. NRC and NNSA agreed with our recommendations to enhance collaboration with other federal agencies, and DHS did not comment on the report.

1.5 Conclusion

According to IAEA, from 1993 to 2013, there were a total of 2,477 confirmed incidents of illicit trafficking in nuclear and radiological materials. The nuclear security summits, spearheaded by the United States, have called greater attention to the importance of strengthening nuclear security worldwide to prevent terrorists from acquiring nuclear materials that could be used in nuclear weapons and radioactive materials that could be used in radiological dispersal devices. The United States has been a leader in promoting nuclear nonproliferation efforts worldwide, and congressional interest in these key issues remains strong. GAO continues to support the Congress by preparing reports and testimonies that help to provide a greater context and understanding of these issues. Current research focuses on a follow-up assessment to the 4-year global nuclear material security initiative as well as a review of key proliferation threats driving current NNSA programs and future program work for the agency. With more than 20 years of experience assessing federal nonproliferation and nuclear security efforts, GAO continues to provide the Congress with timely, objective, fact-based research on these key issues that affect the U.S. government and the American people.

Chapter 2
Neutralizing Radicalized Threat Networks, Disrupting WMD Illicit Traffickers and Targeting Corrupt Facilitators

David M. Luna

Abstract Neutralizing WMD Trafficking Networks and Facilitators: Criminal networks are not only expanding their operations, but they are also diversifying their activities, resulting in a convergence of transnational threats that has evolved to become more complex, volatile, and destabilizing. These networks also threaten U.S. interests by forging alliances with corrupt elements of national governments and using the power and influence of those elements to further their criminal activities. In some cases, national governments exploit these relationships to further their interests to the detriment of the United States.

TOC threatens U.S. interests by taking advantage of failed states or contested spaces; forging alliances with corrupt foreign government officials and some foreign intelligence services; destabilizing political, financial, and security institutions in fragile states; undermining competition in world strategic markets; using cyber technologies and other methods to perpetrate sophisticated frauds; creating the potential for the transfer of weapons of mass destruction (WMD) to terrorists; and expanding narcotrafficking and human and weapons smuggling networks. Terrorists and insurgents increasingly are turning to criminal networks to generate funding and acquire logistical support.

Presentation will outline the threats posed by global illicit trafficking networks including how crime-terror pipelines can be leveraged by transnational organized crime, terrorist groups, and facilitators that provide opportunities for the successful criminal transfer of WMD material to adversaries intent on threatening the welfare and security of the United States and NATO partners.

D.M. Luna (✉)
Senior Director for National Security and Diplomacy, Anti-Crime Programs,
Bureau of International Narcotics and Law Enforcement Affairs (INL),
U.S. Department of State, Washington, DC, USA
e-mail: LunaDM@state.gov

S. Apikyan, D. Diamond (eds.), *Nuclear Threats and Security Challenges*,
NATO Science for Peace and Security Series B: Physics and Biophysics,
DOI 10.1007/978-94-017-9894-5_2

2.1 Illicit Trafficking Networks and WMD Proliferation

We live in a time of insecurity, from deadly infectious diseases such as the Ebola virus, to the proliferation of violent terrorist networks, to the destructive consequences of climate change, and of course, transnational crime, including the horrors that can result from weapons of mass destruction. These are grave threats to our collective security.

Shortly after the September 2014 NATO Summit, held in Wales, President Barack Obama addressed the nation on the threats posed by the Islamic State of Iraq and the Levent (ISIL). The President underscored how we must continue to confront the grave threats posed by terrorism, stating:

> We can't erase every trace of evil from the world, and small groups of killers have the capacity to do great harm. That was the case before 9/11, and that remains true today. And that's why we must remain vigilant as threats emerge.

President Obama also sent a strong message that day to all threat actors and networks around the world who wish to harm Americans and our vital national security interests that those who threaten our country and the safety of Americans will be brought to justice. He said, "This is a core principle: if you threaten America, you will find no safe haven."

A corollary concern is the frightening reality of today's global threat environment consisting of the potential use of chemical, biological, radiological, and nuclear weapons. In some cases, these WMD remain insufficiently safeguarded from theft or other illicit access.

There are still too many terrorists, criminals, and rogue facilitators driven to get access to these WMD; people and networks committed to unleashing these weapons upon scores of innocent people.

Past seizures of weapons-grade nuclear material indicate that such materials continue to circulate on the black market, where they can be bought by criminals and, potentially, transferred to terrorists.

The threat from nuclear terrorism is real. No country is exempt.

Over these next 3 days, distinguished scholars and participants at this conference will address related topics that are of critical importance to NATO and to the international community including the preparedness for nuclear and radiological threats.

These include developing methodologies to identify strategies to measure and reduce the potential risks of trafficking in weapons of mass destruction by illicit actors, and means, methods, and responses to detect, disrupt, and dismantle these threats before they inflict harm on our communities.

Finally, I will advance a novel proposition for the international community to ponder and perhaps begin to view some of today's violent radicalized groups such as ISIL and al-Qaeda affiliates as agents of mass destruction themselves.

2.2 The Origins of a Threat Network

Three years ago, a start-up venture was formed overseas. Much of its early day-to-day operations were small time, simply trying to do what all other nascent businesses try to do: make a name for itself, establish funding and resources, and acquire the talent and equipment necessary to succeed.

In the years since, it has become a global leader in its field, picking up techniques that modern businesses utilize to get ahead, adapt, and leverage market opportunities. Social media and innovative marketing have played a large part of its success—using Facebook and Twitter to amplify its early messages to followers around the world, as well as enable their operations across borders.

As for resources, forget crowdfunding websites—this organization has been able to amass assets in the billions of dollars over the past 3 years. Their adept use of networks, along with some profitable acquisitions, have left them more than adequately funded to take on further expansion.

Most importantly, they are diversifying across markets and borders—they have expanded their portfolio from local affairs to large-scale natural resource management, equipment acquisition and distribution, and dealing in illicit enterprises—from oil bunkering and kidnapping to cigarette smuggling and in trafficking in antiquities.

While there have been reports of some small use of WMD, we are still monitoring plans for larger-scale attacks, as are the citizens of Iraq, Syria, and the rest of the Middle East.

This is the story of the Islamic State of Iraq and the Levant, folks. ISIL.

2.3 Agents of Mass Destruction

A merciless, ideologically-driven terrorist organization bent on creating an Islamic caliphate—from the Levant to Southeast Asia and across Africa—ISIL is in many ways the newest model of a threat network: an international organization that exports fear and exploits profitable opportunities to further its ideological and criminal goals.

Their ideology and violent acts are themselves repulsive as we saw in the litany of horrific crimes against the Yazidis in Iraq. Make no mistake, ISIL would not hesitate to maximize their propensity for further mass violence to the highest degree by unleashing a traditional WMD to murder large numbers of innocent people and continue their march of destruction.

This is why I would venture to add that, in my personal opinion, ISIL, and other terrorist groups, are agents of mass destruction that are killing innocent people, obliterating communities and holy sites, and committing mass atrocities that test the limits of our humanity, such as beheadings, crucifixions, rapes, and other forms of violence.

I do not think that is a stretch, frankly. After all, what is a "weapon of mass destruction"? For certain, when we think of WMD, we think of nuclear, biological, or chemical weapons or devices. But as we witnessed in Boston several years ago, a simple kitchen appliance, a pressure cooker, can also wreak havoc when criminally manipulated.

Suicide bombers can cause mass casualties and instill terror in communities.

Sick individuals that have contracted deadly viruses or pathogens can virtually become bioterror weapons as part of an outbreak.

Nothing today can be ruled out of the realm of possibility. If determined jihadists are willing to blow themselves up, why not do the same using a deadly virus inside their bodies to kill others through transmission?

A new day is upon us where terrorists like ISIL should be viewed as agents of mass destruction.

And ISIL is not alone.

Boko Haram, al-Shaabab, al-Qaeda in the Islamic Maghreb, and others in other regions are similarly using mass fear and destructive violence to inflict murderous crimes in occupied lands as they march to destroy neighboring communities.

Unfortunately terrorist groups are not the only threat networks that the international community has to be concerned about.

Criminal organizations such as the Mexican drug cartels—from the Zetas to the Knights Templar to crime-terror groups such as the D-Company—have terrorized communities using mass violence and intimidation to further their criminal pursuits.

We must acknowledge that these networks of terror and criminality are often no longer just terrorists, and that organized criminal networks aren't just made up of thugs. They are becoming hybrids, ruthless monsters, and agents of mass destruction.

This convergence of terror and criminality has far-reaching consequences and threatens the safety of world citizens especially when threat networks are determined to get their hands on WMD.

I ask you to consider these converging threats through a new prism in your research—how many of today's threat networks such as ISIL are agents of mass destruction, that when converged with a thirst for conventional WMD and a penchant for vicious brutality, become a bigger threat altogether for all nations.

2.4 Illicit Trafficking: A Threat to the Legal Economy and Global Security

Now let me address the dangers of traditional forms of WMD and the perils posed to the international community by illicit trafficking networks.

It is often said that where there is money to be made in illicit markets and the illegal economy, criminals will be very entrepreneurial to oversee and regulate the trading and selling of contraband.

This was certainly true for the A.Q. Khan network in Pakistan as it supplied North Korea and Libya with expertise, technology, and materials to their WMD program.

This brings me to the linkage with another theme of our discussion today.

As underscored earlier, we know that ISIL, Al-Qaeda, and other terrorist groups want to inflict as much damage as they can to innocent communities and their sworn enemies.

In fact, Osama bin Laden in his earlier years with al-Qaeda had stressed to his jihadist followers that it was a "religious duty" to seek and secure WMD as part of their campaigns of terror.

As potential customers and end-users of WMD, ISIL, al-Qaeda, and others, will remain interested in trying to obtain these powerful weapons to maximize catastrophic harm.

To achieve their nefarious terror goals, these groups will likely need to resort to illicit trafficking channels, where corruption, criminals, and black market facilitators come together across supply and demand vectors to obtain WMD.

Their success can only be achieved if we let our guard down.

In the past, we have seen how determined terrorists have obtained mustard and sarin gas, ricin, anthrax, missiles, and other WMD by exploiting corrupt or other vulnerable channels including in countries that have massive stockpiles of WMD or states that are on the verge of great instability and insecurity including Libya, Syria, North Korea, and states of the former Soviet Union.

2.5 An Unholy Trinity: Corruption, Criminality, and Terror

As Dr. Louise Shelley, Director of the Terrorism, Transnational Crime, and Corruption Center (TraCCC), George Mason University, has enlightened us in her recent book "Dirty Entanglements: Corruption Crime, and Terrorism":

> The understanding of financial flows, the role of facilitators from the legitimate economy, and the centrality of particular routes and nodes are crucial to addressing the problem. By focusing only on the crime and terror components, while ignoring the centrality of corruption, it is not possible to effectively address the threat of WMD proliferation, or the possibility of attack. Analysis of the dirty entanglements will be crucial to preventing future attacks.

In a world of convergence, how difficult is our challenge to combat WMD proliferation amidst the entangled webs of corruption, criminality and terrorism?

First, when we realize that nuclear, biological, and radiological materials inhabit many sections of our lives, we begin to comprehend the complex task very quickly.

From hospitals to the smoke detectors in this room, the use of these materials can benefit our everyday lives. However, when used for malicious purposes, nuclear and radioactive materials pose a dangerous and disruptive threat to everyone.

It is because of the potential of this new black market niche that organized criminals have gravitated to seek profit based on demand.

This is basic economics: if there's a product in high demand, then there will be suppliers working to procure and sell it at a profit.

It just so happens that in this case, the product that terrorists and criminals want is closely guarded and extremely dangerous.

In Lyudmila Zaitseva and Kevin Hand's 2003 article "Nuclear Smuggling Chains," they describe the three main types of actors involved in the illegal movement of nuclear material: the suppliers, the intermediaries, and the end-users.

I want to focus particularly on the suppliers and intermediaries—let's call them facilitators, because they are the ones who enable the illicit use of nuclear material, and the ones that we should be most concerned about.

Facilitators can be almost anyone—from organized criminals who traffic hazardous material across borders, to radiology technicians in the hospital, to a corrupt general in a nation's army who wants to bolster his own paycheck.

Anyone who has access to nuclear materials has the potential to be a facilitator or a complicit actor in the illicit trafficking supply chain of WMD proliferation.

Following the break-up of the Soviet Union, we saw nuclear scientists from around the post-Soviet landscape smuggling material out of their labs and attempting to sell it on the black market.

They had suffered a loss of reputation and status in the new post-communist Russia, but most importantly they had lost a degree of financial security.

These scientists believed that there was a profitable market for nuclear materials that they could tap into. This perception holds as well for other facilitators—that there's a huge money-making opportunity out there in trafficking of WMD and hazardous materials.

These are among the challenges we face today.

As long as facilitators believe that there is a market for nuclear and radiological materials or weapons of mass destruction—regardless of whether there actually is—this will place even greater importance on the integrity of those tasked with guarding such materials.

The potential for corruption and blackmail is high: the stakes are higher.

As I mentioned earlier, ISIL was able to obtain a relatively small amount of natural uranium compounds through their rapid territorial expansion. Fortunately the IAEA has assessed that these materials do not present a significant safety, security or nuclear proliferation risk.

The thought, though, that this sadistic group even has access to and could gain access to more material is extremely concerning.

But, ISIL is not alone.

In the past decade or two, we have seen the Japanese cult Aum Shinrikyu use sarin gas on the Tokyo metro. There have also been unsubstantiated reports in the past of Chechen rebels planning to attack Moscow with WMD acquired in Russia's own black markets. And a few years back, there was a case of polonium-210 poisoning in the middle of London that targeted a former Russian spy.

Occasionally, law enforcement agencies across the international community have intercepted several illicit trafficking actors and networks that were attempting to smuggle nuclear and radioactive material through Georgia and other countries of the former Soviet Union.

And of course in places where there is war and instability as there has been recently in Libya, Iraq, and Syria, one has to wonder about whether or not some of the arsenals and stockpiles of WMD have been compromised and fallen into the hands of insurgent and terrorist groups during these conflicts.

In Syria, of course, according to reporting by the United Nations, "chemical weapons were used resulting in the killing of civilians, including many children", and leading UN Secretary General Ban Ki-moon to conclude that a war crime had been committed. Also, as noted earlier, a few weeks ago, it was reported that Islamic State members were engaging in chemical attacks on the battlefield when they dispersed chlorine gas.

This is the threat of WMD that illicit trafficking actors, networks, and facilitators pose to the security of all nations.

2.6 U.S. and International Efforts to Combat TOC and Threat Networks

I've spoken a lot about the "doom and gloom" of what's out there especially at the crossroads of radicalism and the global theater of insecurity and instability.

So what is the U.S. government doing to combat the menace of networks of terror, hostile states, hostile actors, and their criminal facilitators and terrorists who are bent on acquiring WMD to cause mass destruction and violence?

The United States takes these issues very seriously, as articulated in various national security strategies and presidential directives that help inform our efforts to detect and destroy the WMD capabilities and assets of those rogue states and non-state adversaries who threaten the peace and security of the international community by using these weapons, or before they are used.

Implementation of these strategies includes a whole-of-government approach related to military, law enforcement, diplomatic, and intelligence actions to combat terrorism, protect the homeland and our allies through deterrence, disarmament, counter-proliferation, interdiction of WMD materials, technologies, and expertise, and building and strengthening the interdiction capabilities and practices of our partners.

Some of my colleagues from the U.S. government who have greater expertise on WMD proliferation will be sharing more on our overall and current national and international efforts over the next 3 days.

At this time, however, I want to outline some of the diplomatic initiatives, tools, and capacities that the United States is using to address the convergence of threats in many of today's hot spots and to combat the web of corruption and criminality related to illicit trafficking including WMD and to ensure that we keep the world's most destructive weapons out of such destructive hands.

The Department of State has taken extensive bilateral measures to ensure nuclear security around the world, such as strengthening relations with law enforcement agencies in countries in nuclear smuggling hotspots such as Eurasia, the Middle East, and Central Asia.

Multilaterally, we have also heeded the call to action, with regional groups from Africa to Asia and Europe to the Middle East convening to discuss this timely issue.

International organizations such as International Atomic Energy Agency (IAEA), NATO, INTERPOL and the G-7 have been leaders in combating global threats, including the proliferation of WMD, terrorism, and cyber attacks.

Collective action is vital. Efforts such as the Global Initiative to Combat Nuclear Terrorism have helped to strengthen cooperation on information-sharing, conducting joint tabletop exercises, technical assistance, and best practices in areas like nuclear forensics.

The Proliferation Security Initiative (PSI) is an international partnership which aims to stop trafficking of weapons of mass destruction, their delivery systems, and related materials to and from states and non-state actors of proliferation concern.

We must also ensure compliance with international agreements and standards including the Nuclear Non-Proliferation Treaty (NTP), the Chemical Weapons Convention (CWC), and the Biological Weapons Convention (BWC).

In implementing these important treaties and supporting the international organizations that monitor their enforcement, the United States has been working with partners to strengthen the capabilities of our allies to confront threats and prevent unauthorized transfers of WMD and missile technology, expertise, and material; imposing sanctions to discourage such transfers or the willful skirting of these obligations.

On combating cross-border illicit trafficking and criminal networks, the United States works closely with the international community, including INTERPOL, the IAEA, and the UN Office on Drugs and Crime, to strengthen governments' capacities to prosecute those who facilitate the trafficking of nuclear materials and proliferation activities.

We know that effective investigations, prosecutions, and convictions are critical for keeping criminals, terrorists, and other non-state actors from acquiring WMDs and dangerous materials and technologies.

Within the Department of State, INL coordinates closely with other relevant important players including the Bureau of International Security and Nonproliferation (ISN), Bureau of Counterterrorism (CT), Bureau of Arms Control, Verification and Compliance (AVC), and others, to strengthen international cooperation against transnational threats including WMD proliferation and combating illicit networks and corruptive facilitators.

In support of U.S. law enforcement overseas, the Department of State is working with international partners to counter nuclear smuggling and the trafficking of nuclear and radioactive materials. This includes working with the network of Attaches from the Federal Bureau of Investigation (FBI), Homeland Security Investigations (HSI), and other federal agencies, to coordinate with foreign counterparts, as well as international organizations to develop intelligence-based policing operations against suspected smuggling and trafficking networks.

The United States encourages international partners to strengthen capabilities to investigate smuggling networks, locate and remove trafficked material from the black market, and arrest perpetrators.

For example, at the 2012 Nuclear Security Summit in Seoul, 19 countries signed a Statement of Activity and Cooperation to Counter Nuclear Smuggling (CNS), and others announced steps to strengthen counter nuclear smuggling capacities. An updated CNS statement was signed in 2014 at the Nuclear Security Summit in The Hague, and 21 countries outlined steps taken to further strengthen capacities to counter smuggling of these dangerous materials.

Moreover, consistent with United Nations Security Council Resolution 1540 (UNSCR 1540), the United States is helping to advance a global response with committed states to implement measures aimed at preventing non-state actors from acquiring WMD, related materials, and their means of delivery; criminalizing WMD proliferation; strengthening border and export controls (including, for example, financial, transit, and transshipment); combating terrorist financing; and providing technical assistance to states that lack the capacity to implement UNSCR 1540 and properly secure WMD materials.

We look forward to continuing our work on these diplomatic initiatives, partnerships, and cooperative instruments around the globe to prevent, detect and target illicit trafficking of WMD, prohibit proliferation, and provide countermeasures to security sensitive related materials.

On combating illicit networks, recognizing the expanding size, scope, and influence of transnational criminal threats and its impact on U.S. and international security and governance, in July 2011 the White House released the *Strategy to Combat Transnational Organized Crime: Addressing Converging Threats to National Security*.

The TOC Strategy recognizes the threat posed by transnational organized crime and WMD proliferation. The TOC Strategy calls for the U.S. government and our international partners to work together to combat transnational illicit networks, and take that fight to the next level by breaking their corruptive power, attacking their financial underpinnings, stripping them of their illicit wealth, and disrupting their networks.

In support of the TOC Strategy, the U.S. Congress established the Transnational Organized Crime Rewards Program (TOCRP) in order to assist efforts to dismantle transnational criminal organizations and bring their leaders and members to justice.

This new TOC tool complements the Narcotics Rewards Program by offering rewards up to $5 million for information on significant transnational criminal organizations involved in activities beyond drug trafficking, such as human trafficking, money laundering, trafficking in arms, counterfeits and pirated goods, and other illicit trade areas.

Already, we have announced a reward offer of up to $5 million for Li Fangwei, a notorious proliferator of weapons of mass destruction-related technology to Iran. We anticipate that by rewarding informants who provide leads and tips that help hobble transnational criminal organizations, we can protect our citizens, economies, and homeland.

Finally, we must continue to strategically target the unholy trinity of transnational terrorism, crime, and corruption. We must robustly combat corruption across sectors and along the crime-terror continuum.

Over the years, corruption has been a key source of thefts of nuclear and radioactive materials from facilities of origin and related smuggling across borders.

There have been numerous reported cases of insiders stealing small amounts of highly-enriched uranium (HEU), weapons-grade plutonium, and other nuclear and radioactive material to smuggle out of their countries, traffick, and attempts to sell in black markets.

2.7 Peace and Security: Neutralizing WMD Trafficking Networks and Facilitators

In closing, convergence defines the world's threat environment today.

We live in a world in which crimes such as nuclear and radiological material smuggling, terrorism, money laundering, and corruption are often interconnected, with profits from one illicit trade area used to advance further criminal complicity in other areas, and where bad actors will collude to harm communities, directly or indirectly.

As President Barack Obama has underscored: "Halting the spread of weapons of mass destruction is vital to ensuring a peaceful future."

Failure is not an option.

Vigilance, intelligence-sharing, preparedness, and cross-border cooperation can stop the facilitators and the networks from achieving catastrophic ends.

We must be cleared-eyed about the challenges before us and dogged in our efforts to address them.

The United States will continue to remain vigilant in defending the homeland forward against rising threats from transnational terrorism, crime, and corruption including combating nuclear terrorism and enforcing an international non-proliferation system.

My hope is that by the end of this workshop, we can have a productive discussion on the challenges before us and solutions that help the United States, NATO, and other committed international partners and multilateral organizations to more effectively combat the threat posed by the convergence of WMD smuggling and trafficking, illicit networks, and corrupt facilitators.

Through galvanized and sustained action across borders, we can degrade and dismantle today's networks of terror, hostile actors, and corrupt facilitators.

Through collective action and cooperation, we can help to secure WMD so that we can harness our energies to address other grave security threats around the world; keep freedom's light burning to secure the blessings of liberty to those that yearn for peace; and harness a more peaceful world for all of our children free of its deadliest weapons.

Chapter 3
To Pursue an Independent Nuclear Deterrent or Not? Japan's and South Korea's Nuclear Decision Making Models

Emily Cura Saunders and Bryan L. Fearey

Abstract Concerns regarding potential nuclear proliferation activities of South Korea and Japan have increased following recent nuclear tests conducted by North Korea and following regional hegemonic actions by China. Both of these countries have bilateral security agreements with the United States, and are covered under the so-called "nuclear umbrella." This paper poses the question: If these countries enjoy these security guarantees, what factors could contribute to their decision to pursue proliferation activities? This paper reviews some of the relevant literature on nuclear-decision making before proposing that perhaps these models are complementary rather than competing. For the purposes of South Korea and Japan, this paper asserts that there are two contributing factors that could cause either country to engage in proliferation activities: loss of confidence in the U.S. commitment to its extended deterrence obligations and regional security threats. Using case studies, this paper reviews the past proliferation activities by these two countries, and examines the mitigating factors that led to past decisions, and how such similar factors could impact future decisions. This paper uses historical analysis and investigation of contemporary security issues to examine the roles that extended deterrence and regional security have played and may play in the future, in each country's interests in engaging in increased nuclear proliferation-sensitive activities.

E.C. Saunders
Latin American Public Opinion Project, Vanderbilt University, Alamos, New Mexico, USA
e-mail: Emily.c.saunders@vanderbilt.edu

B.L. Fearey (✉)
National Security Office, Los Alamos National Laboratory, Alamos, New Mexico, USA
e-mail: bfearey@lanl.gov

S. Apikyan, D. Diamond (eds.), *Nuclear Threats and Security Challenges*,
NATO Science for Peace and Security Series B: Physics and Biophysics,
DOI 10.1007/978-94-017-9894-5_3

3.1 Introduction

There have been many theories as to why countries pursue nuclear weapons programs. Scholars have speculated that states seek nuclear weapons for a variety of reasons ranging from domestic politics to international security. These models seek to explain the phenomena of why some states develop nuclear weapons and why others abstain. This paper examines these models before turning to case studies. Based on this analysis, this paper proposes an additional way to think about a country's nuclear decision-making.

There are over 30 states currently covered by the United States' extended nuclear deterrent. The crux of this concept is that countries covered by this so-called "nuclear umbrella" need not develop independent nuclear programs because the United States will provide security guarantees in the event of a crisis. Since the formation of the North Atlantic Treaty Organization (NATO), it can be argued that the extended deterrent has been a powerful nonproliferation tool, discouraging a nuclear cascade in several regions. This paper concludes that the perceived "strength" of the U.S. nuclear deterrent plays a role in a given country's nuclear decision-making. Strength here is defined as both numbers and types of nuclear weapons, rhetorical commitments, and military deployments. This definition will be further expanded upon throughout the paper. The two case studies used here are Japan and South Korea. This paper examines what role the strength of the U.S. nuclear deterrent played/plays in the nuclear decision-making of Japan and South Korea.

3.1.1 Existing Framework

In his article, "Why do States Build Nuclear Weapons? Three Models in Search of the Bomb," Scott Sagan challenges the notion that states will proliferate if, and only if, they face an extreme security threat from another international actor. Sagan notes that such a focus fails to accept that "nuclear weapons programs also serve other, more parochial and less obvious objectives."[1] Models of decision-making have been put forth ranging from security to domestic to psychological. This introduction briefly explores some of these models before examining Japan and South Korea as case studies to propose another model, a confidence model. This paper concludes that confidence in the U.S.-provided nuclear umbrella matters, and should be taken into account when looking at a states nuclear decision-making. It should be noted that decision-making criteria do not exist in a vacuum. The model put forth in this paper is meant to supplement existing models: in no case has just one factor led to a state's decision to pursue a nuclear weapons program. Often

---

[1]Sagan [1], 55.

there are a variety of variables, although in some cases one variable may dominate over others—nevertheless, these models must be married to fully explain nuclear decision-making.

The security model is heavily influenced by realist international relations theory. Originally developed by Kenneth Waltz, the theory claims that international actors survive in an anarchic system, and as such they are solely responsible for their safety and security.[2] Sagan sums up the security model with regard to this theory claiming, "because of the enormous destructive power of nuclear weapons, any state that seeks to maintain its national security must balance against any rival state that develops nuclear weapons by gaining access to a nuclear deterrent itself."[3] This is a relatively intuitive notion: if a state feels threatened by a neighboring state's nuclear arsenal, it may be inclined to develop its own, independent deterrent. While intuitive, this model fails to identify the complexity involved in developing a nuclear arsenal, including budgetary, technological, and international constraints.

Etel Solingen moves beyond the security model to explore what domestic factors could play a role in nuclear decision-making. Solingen argues, "Different models of domestic political survival—how leaders seek to gain and maintain power—provide important information regarding nuclear decisions."[4] Solingen concludes that domestic issues are not the only variable, but are largely left out of the security model. She claims that without "taking into account political survival models, one may not properly understand nuclear behavior or estimate the actual effects of the balance of power, international norms and institutions, or democracy."[5] Similarly, this paper concludes that confidence in the U.S. extended nuclear deterrent is but one piece of the puzzle.

Solingen marries international factors with domestic factors. She claims that regional issues matter, explaining the "extent to which regions share congruent orientations toward internationalization (either positive or negative) modifies domestic preferences on nuclear issues."[6] Domestic politics do not exist in a vacuum. Leaders are aware of their place in the international system. With regard to East Asia, Solingen claims the "collective evolution of East Asia toward internationalization reinforced *individual* incentives of leaders to avoid nuclearization to preserve regional stability, foreign investment, and domestic growth."[7] As can be seen today, leaders who chose a path of proliferation are often deemed pariahs, sanctioned economically, and sometimes fought militarily. When a state is outwardly facing, trying to establish a place in the international community, they are less likely to pursue nuclear weapons programs, according to the domestic model. This model is

[2] Waltz [2].

[3] Sagan, "Why do States," 57.

[4] Solingen [3], 39.

[5] Ibid, 44

[6] Ibid, 55.

[7] Ibid, 55.

compelling in that it does not claim to be *the* deciding factor of nuclear decision-making rather it claims that domestic issues must be taken into account in order to create a comprehensive understanding of such decisions.

3.1.2 Coupling Models: Regional Security and Confidence

In 1968 many of the world's nations signed a document that would change the face of nuclear proliferation forever. The preamble of the Nuclear Nonproliferation Treaty (NPT) begins,

> The States concluding this Treaty, hereinafter referred to as the Parties to the Treaty,
> Considering the devastation that would be visited upon all mankind by a nuclear war and the consequent need to make every effort to avert the danger of such a war and to take measures to safeguard the security of peoples,
> Believing that the proliferation of nuclear weapons would seriously enhance the danger of nuclear war,
> In conformity with resolutions of the United Nations General Assembly calling for the conclusion of an agreement on the prevention of wider dissemination of nuclear weapons . . . [8]

The treaty continues to lay out 11 articles that are designed to stem the tide of nuclear proliferation. Forty-one years later, President Obama made another plea to the international community to achieve a world free of nuclear weapons.[9] While this was an impassioned speech on a radical idea, President Obama made sure to assure the United States' allies and assuage the United States' adversaries. He explained, "Make no mistake: As long as these weapons exist, the United States will maintain a safe, secure, and effective arsenal to deter any adversary and guarantee defense to you allies."[10]

The question is how does the United States balance both a reduction in the role and numbers of nuclear weapons in their national defense posture, while still extending the so-called "nuclear umbrella" to more than 30 friends and allies? Japan and South Korea are both allies of the United States and both enjoy bilateral defense agreements with the United States, are there circumstances under which either country may pursue nuclear proliferation related activities?

Two circumstances that could potentially push both Japan and South Korea into considering proliferation related activities are; regional security threats and the loss of confidence in the extended nuclear deterrent provided by the United States. This paper will examine these two issues and determine what, if any, relationship they have with the proliferation related activities of Japan and South Korea. In these cases, the security model is coupled with other variables, one of them being confidence in the U.S. provided extended deterrent.

---

[8]United Nations [4].

[9]Obama [5].

[10]Ibid.

3.2 South Korea

The United States and South Korea have a longstanding relationship. Ever since the North-South Armistice provided a tentative end to the Korean War, the United States has maintained troops on the ground in South Korea. The two countries have coupled this physical presence of both personnel and weapons systems with numerous statements regarding both a general and nuclear security guarantee from the United States.

It is important to understand the initial security relationship between these two states, as well as the historical roots of South Korea's nuclear program.

Signed in 1953 the Mutual Defense Treaty between South Korea and the United States begins,

> Each party recognizes that an armed attack in the Pacific area on either of the Parties in territories now under their respective administrative control, or hereafter recognized by one of the Parties as lawfully brought under the administrative control of the other, would be dangerous to its own peace and safety and declares that it would act to meet the common danger in accordance with its constitutional processes.[11]

This treaty was the first step in formal assurances to the South Korean people that the United States would clearly come to their aid in the event of an act of aggression by another state. It is important to note that there is no explicit language ruling the use of nuclear weapons on behalf of each other in or out.

As with any international treaty, the Mutual Defense Treaty existed in a specific historical moment. As Kang Cho and Joon-Sung Park explain, the treaty "was a joint product of South Korea's desperate effort to keep U.S. forces committed to its defense and the U.S. recognition of Korea's strategic value in fighting the spread of communism in Northeast Asia."[12]

Just three years after signing the Mutual Defense Treaty, the United States reportedly introduced nuclear weapons on the Korean Peninsula, purportedly to maintain a credible security presence, despite a troop drawdown.[13]

3.2.1 *The South Korean Nuclear Option*

After the initial security guarantees of the 1950s, the South Korean-United States relationship was tested when South Korea explored its own nuclear option. It is

[11]"Treaty of Mutual Security Between Japan and the United States of America," Ministry of Foreign Affairs Japan, January 19, 1960. Accessed February 26, 2014, http://www.mofa.go.jp/region/n-america/us/q&a/ref/1.html. "Mutual Defense Treaty Between the United States and the Republic of Korea," *The Avalon Project, Documents in Law, History and Diplomacy*, October 1, 1953. Accessed February 26, 2014, http://avalon.law.yale.edu/20th_century/kor001.asp

[12]Choi and Park [6], p. 374.

[13]Jae-Bong [7].

useful to explore what the determinants were in South Korea's nuclear decision-making and what the response of the United States was with regard to discouraging the nuclear option. Could the United States use this history as a template of how to curb any future South Korean proliferation activities?

In the early 1970s South Korea is believed to have pursued a nuclear weapon option. In 1970, "President Park established the Agency for Defense Development and the Weapons Exploitation Committee to produce or acquire weapons systems and military supplies."[14] Three years later, President Park's "nuclear weapon program began to take concrete form, with reinforcement of nuclear research teams and the effort to acquire nuclear reprocessing plants as well as associated core designs and technologies from France."[15]

With a security guarantee signed 20 years earlier, the proliferation activities of the 1970s beg the question: what were the circumstances in which South Korea felt it needed to attempt to pursue fuel cycle capabilities?

Christopher Hughes from the University of Warwick suggests, "South Korea's perception of declining U.S. implacability in the face of North Korea provocations in the late 1960s, U.S. rapprochement with China in the early 1970s, and U.S. plans to scale back its troop deployments all galvanized President Park to seek nuclear weapons."[16] Richard Bush of the Brookings Institute echoes this sentiment in explaining that the nuclear initiatives "had two drivers: Seoul's increasingly dire threat perception concerning North Korea and an apparent weakening of the U.S. security commitment."[17] Troop deployments are one tangible mechanism in which South Korea has measured the level of commitment from the United States.

3.2.2 *Regional Security Issues Facing South Korea*

What was happening in northeast Asia at the time of these nuclear activities? Could the security model be useful in explaining such activity? The armistice with the North has never been truly stable, and the 1960s and early 1970s were no exception. North Korea attempted to assassinate President Park Chung Hee, and committed several other acts of aggression such as "capturing [the] USS Pueblo, armed guerilla infiltrations, and shooting down the US EC-121 reconnaissance plane."[18] These acts are quite similar to what the international community is seeing today with the most recent provocations from North Korea.

South Korean scholar Seok-soo Lee explains how North Korea's "provocative behavior culminated in an abortive guerrilla attack on the Blue House of presidential

[14] Ibid, 375.

[15] Ibid, 376.

[16] Hughes [8], 94.

[17] Bush [9], 3.

[18] Choi and Park, 375.

residence and infiltration of 120 North Korean special operations forces in 1968. In 1976, [an] ax accident occurred in the Joint Security Area with two casualties of American soldiers."[19] The Blue House raid was an attempt to essentially take over the country, theoretically after the attack, communist agents within South Korea would then mobilize to incite guerilla warfare.

These acts of aggression posed a great threat to South Korea. As a result, South Korea felt insecure and began to explore developing an independent deterrent. The time of these acts of aggression line up with South Koreas nuclear program. However, acts of aggression were not the only thing pushing South Korea to explore the nuclear option; South Korea was beginning to perceive the United States as wavering in their security commitments.

3.2.3 Loss of Confidence

Bush explains that prior to the late 1960s and early 1970s "South Korea was reassured by the mutual defense treaty and the deployment of, first, U.S. troops and then tactical nuclear weapons on the peninsula."[20] It was only when Washington was behaving in a manner that was disconcerting to Seoul, coupled with the security threats from North Korea that South Korea explored the nuclear option.

The aforementioned regional security issues were further exacerbated by the fact that South Korea was quickly losing faith in the United States' capability to protect it. Washington was unable to come to the aid of Seoul during these provocations because it was preoccupied with the war in Vietnam. It is important to note, that the security guarantees are mutual, South Korea assisted the United States with Vietnam. Choi and Park note that South Korea sent "320,000 soldiers to Vietnam over an 8-year period and suffered more than 16,000 casualties including 5,000 killed in action."[21]

Hughes suggests that the "possibility of the alliance dilemma of U.S. abandonment was what formed the prime driver for South Korea's attempt at acquiring nuclear weapons."[22] How does the United States today keep the "alliance dilemma" under control? Would another large conflict that caused the United States to draw down its troop presence cause northeast Asian countries to begin weapons exploration? More importantly, to use the extended nuclear deterrent as a tool to keep potential proliferants from doing so, how can the United States identify and assuage specific fears of abandonment?

---

[19]Seok-Soo Lee, "The Future of Extended Deterrence: A South Korean Perspective," *Fondation por la Recherche Strategique*, No 03/2010, 52.

[20]Bush, 3.

[21]Choi and Park, 375.

[22]Hughes, 94.

After pulling out of the war in Vietnam, President Nixon created a doctrine in which he called for more independence of Asian countries. In 1970, the United States created a plan to withdraw 20,000 troops from South Korea. This was coupled with a long-term plan to completely withdraw by 1975.[23] Troop withdrawal was a real manifestation of the abandonment fears that South Korea felt. To make such a drastic draw down certainly triggered at the very least looking into a South Korean independent nuclear deterrent. Kurt Guthe and Tom Scheber of the National Institute of Public Policy argue, "When doubts about that commitment led South Korea to start a clandestine nuclear weapons program, U.S. assurances, along with other measures, caused Seoul to halt the effort and ratify the NPT."[24]

Lee echoes this sentiment in stating that the South Korean regime under President Park was unconvinced that the extended conventional deterrent provided by the United States was enough. As a result of this, President Park, as mentioned earlier, began to explore nuclear weapons development. The United States applied great pressure to curb this behavior, and once they were successful, the United States "promised to provide the nuclear umbrella to South Korea."[25] This may have created a dangerous precedent as President Park was essentially rewarded for proliferation behaviors. It is important that the United States be cautious and ensures that proliferation activities by South Korea are not rewarded. The United States must be able to anticipate fear of abandonment and understand security threats, before these two variables lead to an increase in proliferation activity.

3.2.4 Contemporary Issues

There is little doubt that the acts of aggression from North Korea to South Korea have been egregious over the past decades. The Korean peninsula rests on a perpetual verge of conflict, with anywhere from 25 to 30,000 troops from the United States standing in the middle. South Korea has previously begun the early steps necessary for a nuclear weapons program at least once in the relatively short history since the armistice.[26] This effort came at a time when South Korea was facing major security infractions from the north, and had lost confidence in the United States' promises to protect.

What are the regional/confidence issues that face South Korea today? Are these extreme enough for South Korea to begin a nuclear weapons program? How can the United States assuage the fear of abandonment and manage the security environment?

---

[23]Choi and Park, 376.

[24]Guthe and Scheber [10], 14.

[25]Lee, 53

[26]Hughes [8], 94.

On October 9, 2006, North Korea conducted an underground nuclear test. This came years after the 1994 nuclear crisis when North Korea threatened to leave the NPT, giving its compulsory 90 days and reneging on day 89, thus opening the door to pull out with one days' notice, which they did in 2003. The test, while egregious, was not so surprising to many in the international community, as North Korea was moving in the direction of nuclear weapons for some time. The United Nations Security Council (UNSC) was quick to act in adopting Resolution 1718 which imposed sanctions on North Korea.[27]

North Korea continued its provocations with another purported test in 2009, and again in 2013. These provocations have been continued by North Korea by doing things like cutting off communications with South Korea and shutting down the Kaesong Industrial Complex.

While the nuclear provocations are certainly alarming, North Korea has also engaged in several conventional aggravations as well. Most notable is perhaps the sinking of one of South Koreas warships, *Cheonan*, which killed 46 soldiers.[28] Six months later, North Korea shelled a South Korean island, Yeonpyeong, killing two marines and injuring many more marines and civilians, not to mention setting several homes ablaze.[29]

Guthe and Scheber explain that as a result of the regional security issues,

> South Korea has indicated interest in exchanges with the United States on a number of topics, such as the military implications of the North Korean nuclear threat, scenarios in which nuclear weapons might play a part, the relationships among the capabilities for extended deterrence (nuclear forces, conventional strike capabilities, and missile defenses), the forces (dual-capable aircraft and strategic nuclear arms) that back the U.S. nuclear guarantee, the ways in which nuclear weapons might be used, and the extent to which South Korea might participate in nuclear related decision-making.[30]

Security issues and provocations led South Korea to wonder if it is time to develop their own nuclear deterrent. *The Los Angeles Times* reported that "Separate opinion polls taken this year by the Asan Institute for Policy Studies and Gallup Korea showed nearly two-thirds of South Koreans in support of nuclear weapons, preferably under their own control."[31] "Preferably under their own control" is a highly significant piece of this polling data. This begs the question, with rising regional tensions, how can the United States bolster their security commitments to South Korea so that they need not develop an independent nuclear deterrent?

There are three issues that have a direct impact on Seoul's level of confidence in the extended nuclear deterrent; the Operational Control (OPCON) of combined

[27]United Nations Security Council [11].

[28]Byun [12].

[29]Powell [13].

[30]Guthe and Scheber, 33.

[31]Barbara Demick [14].

United States/South Korean forces transfer, troop levels and trip line movement, and the 2010 Nuclear Posture Review (NPR). Each of which will be explored next.

Part of the long alliance with South Korea was the establishment of the Combined Forces Command (CFC) in the 1970s. This structure encompasses as plan under which South Korea maintains operational control over its forces during peacetime, however, during wartime, the United States has control over both their forces, as well as South Korea's.[32]

This OPCON arrangement made sense to both the United States and South Korea for the political, security, and regional climate of the 1970s. It was also used as a carrot to prevent the then South Korean administration from further nuclear proliferation activities. In 2007, however, under a much different climate and with the South Korean weapons program in the past, the United States decided to revisit their force commitment to the OPCON and planned for a 2012 transfer, meaning South Korea would have control over its forces in both peacetime and wartime. This suggestion from the United States was met with anxiety on the part of South Korea, and as such the United States has pushed the transfer to 2015 at the request of President Lee Myung-bak.[33] Pushing the date of the transfer is one way the United States can assuage the anxieties of South Korean officials. Ben Rhodes, President Obama's deputy national security advisor for strategic communication has described the delay as a "key signal, particularly given the current state of play on the Korean Peninsula, about the depth of America's commitment to the alliance and to the stability and security of the region."[34] Given the most recent North Korean nuclear test, it is unlikely that the United States will transfer OPCON in 2015. As recently reported, both countries will assess the readiness of such a transfer to the Korean military in 2018, with the OPCON transfer target date in 2020.

The OPCON is one way the United States can demonstrate its commitment to security in the region. Another tangible measurement of this commitment is the levels of United States troops in South Korea. The United States has had troops in South Korea since the Korean War. Troop levels have increased and decreased over time in response to both international and domestic issues. The number of troops in South Korea is often related to help assure South Korea of the United States' security commitment.[35]

While numbers are important, so is placement of the troops. This placement is called the "trip line," and was proposed to be moved away from the Demilitarized Zone (DMZ). Strategically, having United States troops near the DMZ has been a long-term source of comfort for South Korea. If there were a provocation from the north on the border, it would likely guarantee involvement from the United States if troops were killed. If the United States troops were moved south, this could

[32]Guthe and Scheber, 11.

[33]Ibid, 11.

[34]Ben Rhodes, Conference Call Briefing by Ben Rhodes, Mike Froman, Amb. Jeff Bader, Danny Russel, June 26, 2010, White House,

[35]Yong-Soo and Bong-Moon [15].

signal that the South Koreans are alone on the DMZ, and could trigger a sense of abandonment. The United States, however, claimed the move would give it more strategic flexibility in the region, and also ensure the safety of the families of the troops who live in South Korea.

While troop withdrawal and movement along with an OPCON transfer are likely not enough for South Korea to move toward development of their own nuclear deterrent, or to examine the fuel cycle as a hedge option, Seoul does carefully watch these movements with the lens of security assurances. Troop movements/withdrawals coupled with other factors that lead to a loss of confidence in the United States' extended deterrent should be carefully monitored by Washington in terms of how to keep Seoul comfortable with the commitments.

Another consideration for the United States vis-à-vis maintaining a sufficient commitment to its allies is stated policy. The 2010 NPR outlined several ways to meet relevant security challenges. This 2010 NPR, which was made publically available, was carefully reviewed by both United States' allies and adversaries. The 2010 NPR states that the United States can meet security challenges by:

> Reducing the role and numbers of US nuclear weapons—meeting our NPT Article VI obligation to make progress toward nuclear disarmament—we can put ourselves in a much stronger position to persuade our NPT partners to join with us in adopting the measures needed to reinvigorate the nonproliferation regime and secure nuclear materials worldwide.
>
> Maintaining a credible nuclear deterrent and reinforcing regional security architectures with missile defenses and other conventional capabilities, we can reassure our non-nuclear allies and partners worldwide of our security commitments to them and confirm that they do not need nuclear weapons capabilities of their own.
>
> Working to reduce the salience of nuclear weapons in international affairs and moving step-by-step toward eliminating them, we can reverse the growing expectation that we are destined to live in a world with more nuclear armed-states, and decrease incentives for additional countries to hedge against an uncertain future by pursing nuclear options of their own.[36]

This stated policy attempts to meet President Obama's goal of a world without nuclear weapons, while, at the same time, making it clear that the commitments to security partners is unbreakable. The complexity here is that there is a fine line between the nonproliferation regime and security commitments. By reducing the role of nuclear weapons in the United States' posture, and bringing numbers down to fulfill treaty obligations, there is a chance that those under the nuclear umbrella could become dissatisfied with the assurances and seek to proliferate nuclear weapons of their own, thus undermining much of the purpose of the nonproliferation regime.

[36]Nuclear Posture Review.

The relationship between the United States and South Korea has a long and complicated history. The main features of the contemporary relationship with regard to the extended nuclear deterrent and broader defense commitments made by the United States rely both on regional issues and perceived reliability of the nuclear umbrella. The classic security model of nuclear decision-making does not apply to South Korea. They have had significant regional tensions and threats over the years, and have yet to fully pursue a nuclear program. The reasoning for never moving full force with a nuclear program could be that the United States has shown in word and deed that they will protect South Korea in the event of any major security escalation. The times when this commitment has been perceived to be weak coupled with security issues, however, are the times when South Korea has come closest to developing an independent nuclear deterrent. This paper argues that the factors in South Korea's nuclear decision-making move beyond the security model and must take into account the perceived commitment of the United States.

3.3 Japan

The United States and Japan have had a long and, at times, tumultuous relationship. The fact cannot be ignored that Japan is the only country to suffer a nuclear attack by the United States, or any country for that matter. As a result, the troop placements and subsequent security guarantees at the end of World War II are militarily relevant. The first attempt at a formal alliance between the United States and Japan was the 1952 Mutual Assistance Pact, which was built upon in 1960 with the Treaty of Mutual Cooperation.

Article Five of the Treaty of Mutual Cooperation is very similar to the treaty with South Korea. It states, "Each party recognizes that an armed attack against either Party in the territories under the administration of Japan would be dangerous to its own peace and safety and declares that it would act to meet the common danger in accordance with its constitutional provisions and processes."[37]

Article Four states, "The parties will consult together from time to time regarding the implementation of this Treaty, and, at the request of either Party, whenever the security of Japan or the international peace and security of the Far East is threatened."[38] This is an important point, as Japan has been largely unwilling to discuss the presence or absence of nuclear weapons in their waters. Consultations have been rare in the past, however, in recent years, Japanese administrations have been more willing to have frank, but private, conversations about the nature of the United States' extended nuclear deterrent.

For example, the United States issued an NPR in 2001 in which it called of reductions of its nuclear stockpile in some respect similarly to the 2010 NPR.

[37]"Treaty of Mutual Security Between Japan and the United States of America."

[38]Ibid.

Shortly after this intention was made public, according to Guthe and Scheber, "Japanese officials sought clarification from the United States that US nuclear guarantees were still valid and began to ask detailed questions on US plans for the nuclear posture."[39] This diplomatic relationship between the United States and Japan in terms of nuclear policies continues today to be molded and revised based on various security issues. These issues are explored further below.

3.3.1 The Japanese Nuclear Option

Japan's potential nuclear latency has been one of great debate and speculation since the end of the Second World War. There have been many theories as to why Japan would or would not pursue a weapons program, but the two variables identified in this paper, regional security and confidence in the United States' extended deterrent, have strongly influenced this issue. Having been the sole victim of a nuclear attack, Japanese politicians have always taken great care with regard to their rhetoric concerning nuclear weapons. This rhetoric should be carefully monitored by the United States.

Many of Japan's nuclear options can be measured in this highly nuanced political rhetoric. For example, in 1957 under Prime Minister Nobosuke Kishi, the Cabinet Legal Affair Bureau "confirmed that nuclear weapons were not unconstitutional."[40] Domestic pressure and outrage at this claim soon forced Prime Minister Kishi to resign; however, the taboo of talking about Japanese nuclear weapons had been broken.[41]

In the early 1960s Prime Minister Sato went so far as to explicitly tell President Johnson that he was not opposed to exploring a nuclear option for Japan, remarking that, "Japanese public opinion will not permit this at present, but I believe the public, especially the younger generation, can be 'educated.'"[42] Ironically, Prime Minister Sato ended up winning a Nobel Prize for what he deemed the Three Non-Nuclear Principles—no manufacturing, possessing, or presence of nuclear weapons in Japan.[43]

While this change in rhetoric was important, it did not end nuclear exploration in Japan. Several Japanese administrations since Prime Minister Sato have commissioned reports on the feasibility, both scientifically and economically, of developing nuclear weapons. In the context of these administrations the idea of latent capability

---

[39]Guthe and Scheber, 45.

[40]Green and Furukawa [16], 349.

[41]Campbell [17], 221.

[42]Ibid, 222.

[43]Green and Furukawa, 350.

surfaced. In a memorandum written by the director of the Japanese Defense Policy Bureau, Kubo Takyua, he makes this option out to be an insurance plan to keep the United States commitment strong. The memorandum reads,

> If Japan prepares a latent nuclear capability which would enable Japan to develop significant nuclear armament at any time, the United States would be motivated to sustain the Japan-US security system by providing nuclear guarantee to Japan, because otherwise, the US would be afraid of the stability in the international relations triggered by nuclear proliferation.[44]

The commitment of the United States is clearly an issue for Japan. They want to be assured that the commitments are strong, and if not, this memo suggests that they are willing to consider an independent deterrent if need be.

3.3.2 Regional Security Issues Facing Japan

While the aforementioned talk of a Japanese nuclear deterrent spreads across decades, there are certain regional issues that contributed to the nuclear rhetoric. For example, the 1964 Chinese nuclear test was sure to have influenced Prime Minister Sato's rhetoric. Another regional issue that was at play during that time was the rise of the nonproliferation regime that culminated with the creation of the NPT. Japan carefully watched the behavior of other regional countries in this context during that time. Who would sign and who would not sign would directly affect their regional security. An arms build-up in Central Asia could most certainly lead to a ripple effect of arms build-ups in Northeast Asia.

There were several other regional security factors that have also played into the decades long debate among Japanese government officials and citizens. A major issue, as identified by Kurt Campbell in his book, *The Nuclear Tipping Point: Why States Reconsider Their Nuclear Options*, was that "influential Japanese began to fear that with this new nuclear capability, China could make Japan a 'nuclear hostage' in the event of a crisis on the Korean Peninsula, a military conflict in the Taiwan Strait, or an escalation in the Vietnam War."[45] Issues with China continue to be pervasive today.

Another major regional issue was the North Korean nuclear crisis in 1994. Campbell explains, the "crisis on the Korean Peninsula led to speculation in the United States and other countries that if the North Korean situation was not peacefully resolved, the Japanese might reconsider their policy of nuclear abstention."[46] Like the China issue, this is another problem that has continued to prevail today.

---

[44]Ibid, 352.

[45]Campbell, 222.

[46]Ibid, 227.

3.3.3 Loss of Confidence

Japan watches the United States' relationship with other countries, as well as their stated policy and treaty obligations very closely. In the 1980s, the United States negotiated the Intermediate-Range Nuclear Forces Treaty with the Soviet Union. During negotiations, the credibility of the nuclear deterrent was challenged, as the treaty would result in the elimination of shorter-range land-based missiles.

Another source of concern for Japan was the retirement of the Tomahawk Land Attack Missile-Nuclear (TLAM-N). In 2009, the Congressional Commission on the Strategic Posture of the United States explicitly stated,

> In Asia, extended deterrence relies heavily on the deployment of nuclear cruise missiles on some Los Angeles class attack submarines the TLAM-N. This capability will be retired in 2013 unless steps are taken to maintain it. US allies in Asia are not integrated in the same way into nuclear planning and have not been asked to make commitments to deliver systems. In our work as a Commission it has become clear to us that some US allies in Asia would be very concerned by the TLAM-N retirement.[47]

In a letter to then Secretary of State Hillary Clinton, Japan's Foreign Minister Okada responded to the notion that the TLAM-N would be retired. In a highly nuanced letter Okada stated the party line of favoring nuclear disarmament while in the same breath expressing concern stating, "Nevertheless, if TLAM-N is retired, we hope to receive ongoing explanations of your government's deterrence policy, including any impact this might have on extended deterrence for Japan and how this could be supplemented."[48] The United States worked to assuage any anxieties caused by retiring the TLAM-N. In fact, in the 2010 NPR they explicitly stated, "The deterrence and assurance roles of the TLAM-N can be adequately substituted by these other means, and the United States remains committed to providing a credible extended deterrence posture and capability."[49]

3.3.4 Contemporary Issues

There continue to be major security issues facing Japan today. With the drama in the South China Sea, and the 2013 North Korean nuclear test, it is important that the United States makes its commitment clear and strong, so as to avoid Japan's latent option. The main issues facing Japan today are China's budding navy, island disputes, China's economic progress, and the North Korean nuclear and ballistic missile tests. All of these issues fall squarely in the security model of nuclear

---

[47]William J. Perry, chair, "America's Strategic Posture: The Final Report to the Congressional Commission on the Strategic Posture of the United States," US Institute of Peace, 26.

[48]Reported in Jeffrey Lewis, "Japan Hates TLAM-N," *Arms Control Wonk*, January 25, 2010.

[49]Nuclear Posture Review.

decision-making, and therefore it is important to understand that while the United States cannot control regional tensions, it can respond by bolstering confidence in the U.S.-provided nuclear deterrent.

The 2010 Japanese Defense White Paper clearly outlines their concerns with China. One of their largest concerns is with China's multiple missile programs. The White Paper explains, "China possesses various types and ranges of ballistic missiles: intercontinental missiles (ICBM), submarine-launched ballistic missiles (SLBM), intermediate range ballistic missiles/medium range ballistic missiles (IRBM/MRBM), and short range ballistic missiles (SRBM)."[50] Japan is concerned about and well aware of the growing capabilities of China—these missiles could threaten Japan.

Another current issue for Japan is their loss of naval supremacy in the region. The White Paper outlines several provocations,

> Advancements to the Pacific Ocean by Chinese naval surface vessels have also been confirmed. For example, in October 2008, four Chinese naval vessels, including Soveremenny-class destroyer, passed through Tsugaru Strait and sailed south to the Pacific Ocean to circle Japan. In November 2008, four naval vessels, including top-of-the line Luzhou-class destroyer, passed between Okinawa Island and Miyako Island and headed to the waters northeast of Okintori Island before engaging in apparent drills. In March 2010, six naval vessels, including a Luzhou-class destroyer, passed between Okinawa Island and Miyako Island and headed to the Pacific Ocean. These vessels were reported to have advanced to the South China Sea.[51]

Issues regarding China's increasing maritime supremacy continue. Recent provocations over the Senkaku Islands are but one example of rising tensions in the region. *The Financial Times* reported that in February 2013, "Tokyo accused the Chinese navy of 'painting' a Japanese warship with weapons-targeting radar."[52] This situation could escalate quickly given that the Japanese have been feeling threatened for years by China, and Prime Minister Shinzo Abe is considered to be far more conservative on military matters than his predecessors. The same article notes that he "has ordered the first increase in Japan's military budget in 11 years, and set a longer-term goal of amending Japan's constitution to loosen constraints on its armed forces."[53]

The United States has made clear their intention to stand with Japan on these issues, however. In April of 2013 Secretary of State John Kerry met with various high-ranking officials in both the Japanese and South Korean governments. On the issue of the Senkaku Islands, "Kerry gave assurances that the U.S. position

[50] Ministry of Defense [18], 58.

[51] Ibid, 61.

[52] Soble [19].

[53] Ibid.

is also unchanged, that the Japan-U.S. security treaty would be applied to foreign attacks on the Senkakus, which are controlled by Japan but claimed by China and Taiwan."[54]

Along with concerns over China's naval program, of particular concern is China's nuclear weapons posture. Since the Chinese program began, they have had only first strike capability, using their nuclear weapons as a deterrent. There is concern in Japan that China could consider a second-strike capability, bolstering their program and posture.[55] Japan is so concerned with this issue that in 2010, Japanese Foreign Minister Okada broke from the usual taboo of discussing nuclear weapons. During regional talks, Okada told his Chinese counterpart, Jiechi Yang, "Amongst the P5, it is only China which is increasing its nuclear arsenal... Therefore I would like to request the Chinese government either reduce the number of nuclear arsenals or at least commit ourselves not to increase its nuclear arsenal from the current level."[56]

China is not the only country Japan worries about in the region. North Korea has continually provoked Japan by testing nuclear weapons and launching ballistic missiles into the Sea of Japan. The 2013 Japanese White Paper clearly identifies concern over North Korea's nuclear program. The report details the February 2013 North Korean nuclear test claiming

> Moreover there is a possibility that the country is developing nuclear weapons using highly enriched uranium. Considered in conjunction with its efforts to enhance its ballistic missile capability, the nuclear tests by North Korea pose a significant threat to Japan's security, and they are significantly detrimental to peace and stability in Northeast Asia and the international community. Therefore, they are absolutely unacceptable. Further actions by North Korea continue to be unpredictable and Japan needs to pay utmost attention to them.[57]

Like the United States-South Korea relationship, the Japan-United States relationship is complicated, and at times, convergent. This relationship is both politically and strategically important for both sides. There are many sources of tension that could lead Japan to lose faith in the extended nuclear deterrent provided by the United States, namely, United States-Sino relations, how the United States deals with the North Korean situation, and the 2010 NPR.

A major security concern for Japan is a rising China, both in its military strength and economic policies. The Japanese worry that their economic edge of the 1980s has been in decline due to competition with China, and as a result, China is becoming competitive vis-à-vis regional economic dominance. The Center for Strategic and International Studies hosted a *Third US-Japan Strategic Dialogue* forum in which they note, "Japan is seriously studying the American strategic

---

[54] Aoki and Yoshida [20].

[55] Green and Furukawa, 354.

[56] *Japan, China Spat over Nuclear arsenal.*

[57] Ministry of Defense, Japan, *White Paper 2013*, Section Two: Security Environment in the Vicinity of Japan, 3. http://www.mod.go.jp/e/publ/w_paper/pdf/2013/07_Part1_Chapter0_Sec2.pdf

posture in this region. Will the US choose Japan or China as a strategic partner? Will it choose neither? Unless the US chooses to promote the Japan-US relationship to a level like that of the Anglo-American special relationship, Japan may withdraw from the alliance and the US may find itself an isolated power in the Pacific."[58] This relationship needs to be carefully considered when thinking about Japan's potential nuclear.

Similarly, Japan watches how the United States deals with North Korea. Japanese officials want to know how the United States will stem the tide of proliferation in the region. The United States has a long history of assurances when it comes to Japanese anxieties over nuclear posture. A more recent development is that Japan has asked to be more involved in the process of nuclear posture planning. *The Japan Times* noted that "Two days after the latest nuclear test by the North, U.S. President Barack Obama told Prime Minister Shinzo Abe in a telephone meeting that there would be no change whatsoever in America's commitment to defend Japan, including nuclear deterrence through its nuclear umbrella over Japan."[59] The United States needs to continually express its intent to provide Japan with security in both words and deeds. It is important that the United States continually involves Japan in discussions concerning the nuclear umbrella.

Japan finds itself on the same tightrope the United States finds itself on, specifically how to applaud/pursue reductions of and emphasis on nuclear weapons while simultaneously maintaining a safe, secure, and effective nuclear deterrent. Japan was apparently sufficiently satisfied with the final draft of the 2010 NPR and their consultations beforehand. However, Japan was insistent on further discussions addressing more willing to talk of their specific concerns, holding high-level meetings to discuss the details and implications of the 2010 NPR. There were, however, still a few points of contention.

One of the largest issues between Japan and the United States on the NPR is the different threat priorities. While Japan is not impervious or ignorant towards terrorism, they see state actors as their main threats, whereas the NPR identified nuclear terrorism as the highest priority, specifically "preventing nuclear proliferation and nuclear terrorism."[60] Japan, on the other hand, sees China and North Korea as its biggest threats, not terrorism. In the Pacific Forum dialogue at the Center for Strategic and International Studies, "Japanese participants have warned of a potential problem should the United States not pay sufficient attention to traditional state-based nuclear security threats."[61]

Another point of contention was how the 2010 NPR dealt with China. The NPR stated,

---

[58]Pacific Forum CSIS [21].

[59]"Nuclear Arms Card for Japan," *The Japan Times*, April 29, 2013. Accessed July 15, 2013: http://www.japantimes.co.jp/opinion/2013/04/29/commentary/nuclear-arms-card-for-japan/#.Ud8x-T7wK50

[60]Nuclear Posture Review.

[61]Nuclear Posture Review.

> The United States and China are increasingly interdependent and their shared responsibilities for addressing global security threats, such as WMD proliferation and terrorism, are growing. The United States welcomes a strong, prosperous, and successful China that plays a greater role in supporting international rules, norms, and institutions.[62]

This is in stark contrast to the stance as articulated in from Japan's White Papers (*vide supra*).

This less-than-confrontational language in the 2010 NPR further increases Japan's anxieties about the U.S.-Sino relationship. While the fear of abandonment may not be analytically founded, the fear remains that perhaps the United States will look for different allies in the region. Should Japan believe that the United States and China were becoming closer, would they be inclined to either look for security guarantees elsewhere, or move toward and independent deterrent?

While these sources of tension exist, Japan is nevertheless very proud of its partnership with the United States, and vice-a-versa. The aforementioned 2013 White Paper highlighted the US-Japan partnership. It explained the "peace and security of Japan is ensured through developing seamless defense measures by coupling Japan's own defense capabilities with the Japan-U.S. Security Arrangements."[63]

Japan, like South Korea, has never actually pursued a nuclear weapons program. They have, however, like South Korea had major security threats and tensions. The security model does not hold true in this case. Japan is surrounded by two nuclear weapons states, each of which has continually provoked Japan, and yet, while some of these threats are quite provocative, Japan has yet to respond with direct proliferation activities. Japan does, however, closely monitor its relationship with the United States, and when regional threats are coupled with a perceived wavering of U.S. support and commitment, Japan is more likely to hint at an independent deterrent.

3.4 Conclusion

Since the formation of the North Atlantic Treaty Organization (NATO), the extended deterrent of the United States has been a powerful nonproliferation tool, discouraging a nuclear cascade in several regions. The analysis performed and discussed in this paper concludes that the perceived "strength" of the U.S. nuclear deterrent (numbers, types, commitments, and military deployments) plays a critical role in a given country's nuclear decision-making process. The paper examined in detail the roles that U.S. nuclear deterrent has played in the nuclear decision-making of Japan and South Korea.

---

[62] Nuclear Posture Review.

[63] Ministry of Defense, Japan, *White Paper 2013*, Part 2, Section 3, p. 100. Accessed February 26, 2014/ http://www.mod.go.jp/e/publ/w_paper/pdf/2013/21_Part2_Chapter1_Sec1.pdf

Models on decision-making have ranged from security to domestic to psychological. This paper assesses these models and examines Japan and South Korea, as case studies, and finds that an additional model, a confidence model is needed. It concludes that *confidence* in the United States nuclear umbrella matters, and should be taken into account when looking at a state's nuclear decision-making. This confidence model put forth here supplements existing models, as no single factor would likely lead to a state's decision to pursue nuclear weapons program. All of these models should be married together to fully explain a country's nuclear decision-making process.

Disclaimer Remarks expressed in this article are solely the authors' own and do not represent those of the Los Alamos National Laboratory, the National Nuclear Security Administration, the Department of Energy or any other U.S. government agency.

References

1. Sagan S (1996/1997) Why do States build nuclear weapons? Three models in search of a bomb. Int Secur 21(3): 55
2. Waltz KN (2008) Realism and international politics. Routledge, Chapman, and Hall, London
3. Solingen E (2010) Domestic models of political survival: why some do and others don't (Proliferate). In: Potter WC, Mukhatzhanova G (eds) Forecasting nuclear proliferation in the 21st century: the role of theory. Stanford University Press, Stanford, p 39
4. United Nations (1968) Treaty on the nonproliferation of nuclear weapons. United Nations, New York
5. Obama B (2009) Remarks by President Obama. (speech Prague, Czech Republic, April 5, 2009), The White House http://www.whitehouse.gov/the_press_office/Remarks-By-President-Barack-Obama-In-Prague-As-Delivered
6. Choi K, Park J-S (2008) South Korea: fears of abandonment and entrapment. In: Algappa M (ed) The long shadow: nuclear weapons and security in 21st century Asia. Stanford University Press, Stanford, p 374
7. Jae-Bong L (2009) U.S. deployment of nuclear weapons in 1950s South Korea & North Korea's nuclear development: toward denuclearization of the Korean Peninsula. Asia-Pacific J Jpn Focus, February 17, 2009. http://www.japanfocus.org/-lee-jae_bong/3053. Accessed 26 Feb 2014
8. Hughes CW (2007) North Korea's nuclear weapons: implications for the nuclear ambitions of Japan, South Korea, and Taiwan. Asia Policy 3:94
9. Bush RC (2011) The US Policy of Extended Deterrence in East Asia: history, current views, and implications. Foreign Policy as Brookings, Arms Control Series, Paper 5, p 3
10. Guthe K, Scheber T (2011) Assuring South Korea and Japan as the role and number of US nuclear weapons are reduced. Defense Threat Reduction Agency Advanced Systems and Concepts Office, Report Number ASCO 2011 003, p 14
11. United Nations Security Council (2006) Resolution 1718 Nonproliferation/Democratic People's Republic of Korea, 14 October 2006, (S/RES/1718). http://www.un.org/ga/search/view_doc.asp?symbol=S/RES/1718%282006%29. Accessed 15 July 2013
12. Byun S-W (2010) North Korea's provocations and their impact on Northeast Asian regional security. Center for US-Korea Policy
13. Powell B (2010) Behind the Koreas' artillery fire: Kim's succession. *Time*, November 23, 2010
14. Demick B (2003) More South Korean's support developing nuclear weapons. The Los Angeles Times, May 18. http://articles.latimes.com/2013/may/18/world/la-fg-south-korea-nuclear-20130519

15. Yong-Soo J, Bong-Moon K (2014) Terms of wartime OPCON discussed. Korea Joongang Daily, September 14, 2014. http://koreajoongangdaily.joins.com/news/article/article.aspx?aid=2995120
16. Green M, Furukawa K (2008) Japan: new nuclear realism. In: Algappa M (ed) The long shadow: nuclear weapons in 21st century Asia. Stanford University Press, Stanford, p 349
17. Campbell KM (2004) The nuclear tipping point: why states reconsider their nuclear choices. Brookings Institution, Washington, DC, p 221
18. Ministry of Defense (2010) The Defense of Japan, p 58
19. Soble J (2013) Japan criticises China over maritime disputes in white paper. The Financial Times, July 9, 2013. http://www.ft.com/cms/s/0/d676d58e-e84a-11e2-9aad-00144feabdc0.html#axzz2YmUvGnPI. Accessed 15 July 2013
20. Aoki M, Yoshida R (2013) U.S. will protect Allies, Kerry Assures Tokyo. The Japan Times, April 15, 2013. http://www.japantimes.co.jp/news/2013/04/15/national/u-s-will-protect-allies-kerry-assures-tokyo/#.Ud8-2z7wK50. Accessed 13 July 2013
21. Pacific Forum CSIS (2001) TMD and Japan: the first state of integrating TMD and NMD into Alliance Missile Defense. Issues and Insights 1.

Chapter 4
Overview of the Cooperative Projects Implemented by the European Commission Joint Research Centre in the Nuclear Security Area Outside Europe

Paolo Peerani, V. Berthou, W. Janssens, and K. Mayer

Abstract This paper starts with an historical overview of the different outreach initiatives and correspondent funding schemes supported by the European Commission in the field of nuclear security. Then it analyses the results of past collaboration under the TACIS and IfS cooperative projects, presents the status of the follow-up programme being implemented and gives an overview of the JRC expertise which can be utilized in a new nuclear safeguards and security programme, as well as under the Centre of Excellence initiative.

4.1 Introduction

The European Union has adopted a series of policy decisions which focus on enhancing nuclear security and non-proliferation world-wide. These are closely connected to the European Union Common Foreign Security Policy and follow the international developments in this area (such as the "EU Strategy against proliferation of Weapons of Mass Destruction 2003"; "Instrument for Stability, 2006, CBRN risk mitigation component"; "Major lines to combat the proliferation of WMD", issued end of 2008, EU CBRN Action Plan and the review of the EU Dual Use Regulation 428/2009).

In line with these policies a significant number of outreach, support and capacity building projects have been funded by the EU in the area of nuclear safeguards, non-proliferation and nuclear security. This paper will focus mainly on those projects, executed by the Joint Research Centre of the European Commission, and will present results, lessons learned and future plans.

P. Peerani (✉) • V. Berthou • W. Janssens
Institute for Transuranium Elements, Joint Research Centre, European Commission, Ispra (Italy)
e-mail: peerani.paolo@jrc.ec.europa.eu

K. Mayer
Institute for Transuranium Elements, Joint Research Centre, European Commission, Karlsruhe (Germany)

S. Apikyan, D. Diamond (eds.), *Nuclear Threats and Security Challenges*, NATO Science for Peace and Security Series B: Physics and Biophysics,
DOI 10.1007/978-94-017-9894-5_4

In the early 1990s, following the breakdown of the former Soviet Union, the European Commission (EC) initiated a Technical Assistance to the Commonwealth of Independent States, the TACIS support program. In the initial phase, essentially, nuclear safety projects were funded under the TACIS program. From 1994, projects related to nuclear safeguards were included in the TACIS program. The significant experience that the JRC has built up in measuring and controlling nuclear material through its involvement in the safeguards area has been made available and transferred to CIS countries through dedicated projects carried out in the framework of the TACIS program. The TACIS program started in 1992 and ended in 2006. After 10 years, the successful EC-CIS cooperation has evolved from a demand-driven to a discussion-driven relationship in areas of mutual interest and benefit. The follow-up program, taking into account new international threats, included combating of illicit trafficking while sustaining past initiatives within an enlarged international cooperation. These cooperative projects were launched in 2005 with initial funding from TACIS and continued under two instruments: the Instrument for Stability (IfS) launched in 2006 and dedicated to nuclear security and the Instrument for Nuclear Safety Cooperation (INSC) started in 2007 under which nuclear safeguards projects are still funded. In the field of nuclear security, the Instrument for Stability extends the geographical scope of the support worldwide. South East Asia, the Mediterranean Basin and Central Asia became beneficiary countries where support projects are currently implemented. The JRC co-coordinates its efforts with other ongoing international activities and major international support programs under the Border Monitoring Working Group to avoid duplication, to identify gaps and to provide an integrated response to the illicit trafficking of nuclear and other radioactive materials.

Under the Instrument for Stability, a new initiative has been launched and financed: the Centre of Excellence. This innovative EU approach, which needs to rely on the full collaboration and input from EU Member States (with a large variety of stakeholders in the MS, including Ministries, Competent Authorities, R&D organizations, academia, industries and NGO's) seeks to reduce the CBRN risk outside Europe (and thus increase the stability in those regions with a return of benefit for the EU security) and to enhance the preparedness to CBRN threats to limit the potential consequences. The CoE concept focuses on the establishment of sustainable and regional approaches, co-owned by the local partners, to enhance the institutional capabilities of the region, countries and all local stakeholders to be better prepared for the prevention, early recognition and/or detection and the structured response to a CBRN threat (no matter whether this comes from a terrorist attack, a natural cause or an industrial hazard). The implementation of the collaborative projects, which are currently being set-up with the partner countries in eight regions world-wide, will be done in close coordination and collaboration with the IAEA (especially in view of the nuclear security support centres) and with other major world-players (like USA, Japan, Australia, South Korea, China etc.) and also taking into account already existing (often bilateral) EU Member States programmes.

This paper analyses the results of past collaboration under the TACIS and IfS cooperative projects, presents the status of the follow-up programme being implemented and gives an overview of the JRC expertise which can be utilized in a new nuclear safeguards and security programme, as well as under the Centre of Excellence initiative.

4.2 The TACIS Program (1992–2006)

The TACIS program (Technical Assistance to the Community of Independent States) was launched by the European Commission in 1992 as a cooperative initiative having the main purpose to finance projects aiming to support the Russian Federation and the other CIS countries in the field of nuclear safety, and in particular for post-Chernobyl remediation actions. Starting in 1995 the program has broadened its scope and included also projects aiming to build a safeguards culture and to enhance capacities in Nuclear Material Accountancy and Control in those countries. Finally from the first years of the new century, the increased concern on terrorism has further focused several projects in the direction of the fight against the illicit trafficking of nuclear material and border control.

Due to its historical competence and expertise in developing methodology and technology in support to nuclear safeguards inspectorates, EURATOM and IAEA, JRC has been directly tasked by the European Commission to implement the projects in the field of nuclear safeguards and security. As an example of the type of projects, Table 4.1 reports the actions performed in the last batch of projects funded in the FY 2005.

The projects more specifically dedicated to border control and fight against illicit trafficking were characterized by a large component of installation of equipment to detect and respond to the illicit trafficking of radioactive and nuclear materials. The typical structure of these projects was:

- Analysis of the nuclear security framework and of the existing procedures at borders in the country
- Analysis and prioritization of the need of equipment
- Development of technical specifications
- Procurement and installation of equipment
- Support to development and/or validation of standard operating procedures
- Training and qualification of personnel

4.3 The Instrument for Stability (2007–2013)

With the new cooperation scheme launched in 2007, the Instrument for Stability, the geographic area of the potential beneficiary countries has geographically enlarged even though initially the projects on nuclear security started in a limited set of

Table 4.1 TACIS projects implemented by JRC in the FY 2005 budget

	Project	Beneficiary
A	Improvement of accountancy and control of hold-up and waste in RT-1 plant at Mayak	Russia
B	Establishment of testing laboratory at VNIIA for certification of NMAC instruments	Russia
C	Development and introduction of modern sealing devices at Minatom's enterprises	Russia
D	Analytical and metrological support on NMAC	Russia
E	Nuclear material accountancy and control (NMAC) applied to naval spent fuel in Andreeva Bay	Russia
F	Implementation of measures to combat illicit trafficking of radioactive and nuclear material	Regional
G	Containment/surveillance system for RBMK spent fuel storage on Kursk NPP	Russia
H	Ukrainian border crossing station (measures of fight against illicit trafficking of nuclear and radioactive material)	Ukraine
I	Armenian border crossing station (measures of fight against illicit trafficking of nuclear and radioactive material)	Armenia
J	Adaptation and commissioning of a computerized NMA system in Armenian NPP Medzamor	Armenia
K	Enhancing the capability for analysis of seized nuclear material and radioactive substances by the main organization of Ukraine	Ukraine
L	Enhancing the capability for analysis of seized nuclear material and radioactive substances by the main organization of Kazakhstan	Kazakhstan
M	Automated data analysis and interpretation for near real time accountancy at the Ulba metallurgical plant	Kazakhstan
N	Sustainability of Ural-Siberian Methodological Research Center	Russia

prioritized regions: South-East and Central Asia, Mediterranean Basin and former Soviet Union.

The structure and typology of projects has also evolved in the sense that the component of equipment procurement has been reduced in favor of actions more directed towards capacity building, improving the national legislation, collaboration among national authorities, preparedness and establishing response procedures.

Table 4.2 reports some examples of typical projects implemented by JRC under the IfS in the recent years.

4.4 The CBRN Centres of Excellence

The current initiative has introduced a total innovation in the way to perform outreach. The classical top-down approach, where the donor was directly negotiating with a country the content of the project within the boundary of the initially

Table 4.2 IfS projects implemented by JRC

Project	Beneficiary
Belarussian border crossing stations	Belarus
Border monitoring in Tajikistan	Tajikistan
Border monitoring in Uzbekistan	Uzbekistan
Border monitoring in Georgia	Georgia
Border management in the Mediterranean Basin	Algeria
Border management in the Mediterranean Basin	Morocco
Border management in South East Asia	Regional: Cambodia, Philippines, Lao PDR and Thailand
Border monitoring in Dem. Republic of Congo	DR Congo

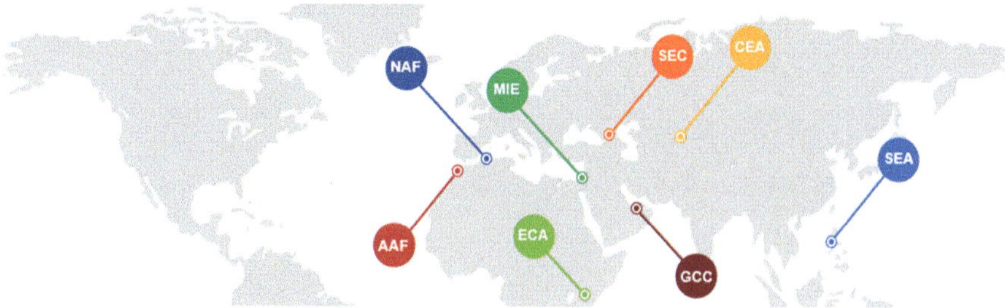

Fig. 4.1 Regional secretariats of the CBRN Centres of Excellence

approved portfolio of activities and budget, the CBRN-CoE initiative reverses the scheme to a sort of top-driven bottom-up process. It is now the beneficiary country that is guided through a self-assessment of its own needs, via a Needs Assessment Questionnaire (NAQ) followed by a National Action Plan describing the priority areas, to propose projects to be funded by the EU via the CBRN CoEs. The projects are further implemented by EU member states expert organisations.

The JRC has been tasked to technically support the CBRN CoE initiative and follow the whole steps of the process, via the organisation of needs assessment meetings, on demand of partner country, support the preparation of national action plan through the organisation of workshop, also on demand of the partner country, draft the terms of reference of newly approved projects, technically evaluate the implemented projects and make an impact assessment.

The proposals are coordinated at regional level by eight regional secretariats (Fig. 4.1):

- African Atlantic Façade (Regional Secretariat in Rabat, Morocco)
- Central Asia (Regional Secretariat in Tashkent, Uzbekistan)
- Eastern and Central Africa (Regional Secretariat in Nairobi, Kenya)
- Gulf Cooperation Council Countries (Regional Secretariat in Abu Dhabi, United Arab Emirates)
- Middle East (Regional Secretariat in Amman, Jordan)

- North Africa (Regional Secretariat in Algiers, Algeria)
- South East Asia (Regional Secretariat in Manila, the Philippines)
- South East Europe, Southern Caucasus, Moldova and Ukraine (Regional Secretariat in Tbilisi, Georgia)

The 47 current projects approved at this date can be found at the following website: http://www.cbrn-coe.eu/Projects.aspx.

Chapter 5
Proliferation Resistance and Physical Protection (PR&PP) Evaluation Methodology and Applications

Robert A. Bari

Abstract This paper presents a summary of a program within the Generation IV International Forum on the evaluation methodology for proliferation resistance and physical protection (PR&PP) of advanced nuclear energy systems (NESs). For a proposed NES design, the methodology defines a set of challenges, analyzes system response to these challenges, and assesses outcomes. The challenges to the NESs are the threats posed by potential actors (proliferant States or sub-national adversaries). The characteristics of NESs, both technical and institutional, are used to evaluate the response of the system and to determine its resistance against proliferation threats and robustness against sabotage and terrorism threats. The outcomes of the system response are expressed in terms of a set of measures, which are the high-level PR&PP characteristics of the NES.

5.1 Introduction

5.1.1 Definitions of PR&PP for NESs

The definitions of Proliferation Resistance and Physical Protection [1] that have been adapted for nuclear energy systems (NESs) by the Generation IV International Forum (GIF) are as follows:

Proliferation resistance is that characteristic of an NES that impedes the diversion or undeclared production of nuclear material or misuse of technology by the Host State seeking to acquire nuclear weapons or other nuclear explosive devices.

Physical protection (robustness) is that characteristic of an NES that impedes the theft of materials suitable for nuclear explosives or radiation dispersal devices

R.A. Bari (✉)
Brookhaven National Laboratory, Upton, NY, USA
e-mail: bari@bnl.gov

S. Apikyan, D. Diamond (eds.), *Nuclear Threats and Security Challenges*,
NATO Science for Peace and Security Series B: Physics and Biophysics,
DOI 10.1007/978-94-017-9894-5_5

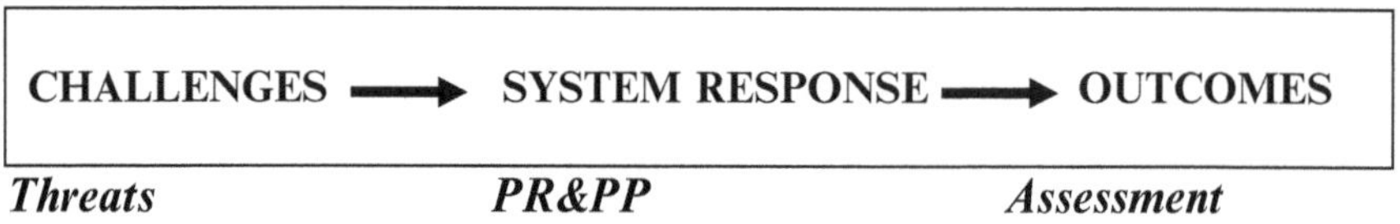

Fig. 5.1 Basic framework for the PR&PP evaluation methodology

(RDDs) and the sabotage of facilities and transportation by sub-national entities and other non-Host State adversaries.

Figure 5.1 illustrates the methodological approach at its most basic. For a given system, analysts define a set of *challenges*, analyze *system response* to these challenges, and assess *outcomes*.

The challenges to the NESs are the threats posed by potential proliferant States and by sub-national adversaries. The technical and institutional characteristics of the NESs are used to evaluate the response of the system and determine its *resistance* to proliferation threats and *robustness* against sabotage and terrorism threats. The outcomes of the system response are expressed in terms of PR&PP *measures* and assessed.

The evaluation methodology assumes that an NES has been at least conceptualized or designed, including both the intrinsic and extrinsic protective features of the system. Intrinsic features include the physical and engineering aspects of the system; extrinsic features include institutional aspects, such as safeguards and external barriers. A major thrust of the PR&PP evaluation is to elucidate the interactions between the intrinsic and the extrinsic features, study their interplay, and then guide the path toward an optimized design.

The structure for the PR&PP evaluation can be applied to the entire fuel cycle or to specific elements of the chosen fuel cycle (reactor, front-end, or back-end of the particular fuel cycle under consideration). The methodology is organized as a *progressive* approach to allow evaluations to become more detailed and more representative as system design progresses. PR&PP evaluations should be performed at the earliest stages of design when flow diagrams are first developed in order to systematically integrate proliferation resistance and physical protection robustness into the designs of NESs along with the other possible high-level technology goals, such as safety and reliability, and economics. This approach provides early, useful feedback to designers, program policy makers, and external stakeholders from basic process selection (e.g., recycling process and type of fuel), to detailed layout of equipment and structures, to facility demonstration testing.

5.1.2 Some Safeguards and Security Considerations

The currently proposed GIF NESs may have new design features and technologies that could require new tools and measures for safeguards and security. Some international safeguards and national security considerations for advanced NESs may be different from those for existing nuclear systems.

International safeguards typically verify the operator's declaration of activities with nuclear material. These declarations address the receipts, shipments, storage, movement, and production of nuclear material. Inspections depend on the material type and whether the material is irradiated. The IAEA state level approach [2] will in addition take into account the technical capabilities of the state including the possible existence of other nuclear activities (including commercial or academic R&D) and the location of the facilities.

The following considerations may pertain to physical security of an NES [3]:

- Location and configuration of vital components so that gaining access to these components is extremely difficult and time consuming for an intruder
- Location and configuration of critical safety systems so that there is not a capability to destroy multiple vital components from a single location
- Incorporation of multiple layers of delay barriers against intruders and minimize the number of access points to areas containing vital assets

Consideration should also be given to physical security system designs options which minimize human involvement in security events, minimize impact of necessary future system modifications, and maximize adversary delay times.

5.2 Some Frequently Asked Questions (FAQ) About PR&PP

Recently, the PR&PP Working Group developed a list of frequently asked questions about the methodology and its applications. These questions and the corresponding answers are given here. They can also be found at: https://www.gen-4.org/gif/jcms/c_44998/faq-on-proliferation-resistance-and-physical-protection.

1. What is the PR&PP evaluation approach?

   The PR&PP methodology is a systematic and comprehensive tool for assessing and optimizing, at all stages of design, the level of proliferation resistance and physical protection of a nuclear energy system, or components thereof. It is a "pathways evaluation" approach which can account for a full range of hypothetical proliferation or terrorism scenarios (including diversion, misuse, clandestine operation, sabotage, and theft), and compute their impact against a set of high level measures.
2. Why was the PR&PP methodology developed?

   The PR&PP methodology was developed to address one of the four goals identified for future nuclear energy systems in the 2002 Generation IV Roadmap (i.e., next-generation power reactor designs that will see commercial deployment beyond 2030); "Generation IV nuclear energy systems will increase the assurance that they are a very unattractive (i.e., present significant barriers) and the least desirable route for diversion (i.e., removal by a State from a declared safeguarded facility, or used to produce undeclared nuclear material) or theft of weapons-usable materials, and provide increased physical protection against acts of terrorism."

3. What is the level of effort needed to perform a PR&PP evaluation?

   The level of effort depends on the stage of design, the range of challenges evaluated, and the needs of the user performing the evaluation. The methodology is adaptable to differing needs. It can involve a single PR&PP expert with subject matter expert support from design staff (for a scoping study), or a team, requiring a few staff-months to a few staff-years.

4. What is the time needed to perform a PR&PP evaluation?

   The time requirement can be as little as a few weeks of work for a scoping study that evaluates the system response a small number of representative PR or PP challenges, to a year or more of work to evaluate response to a comprehensive spectrum of challenges.

5. What is the form of the results?

   The results take the form of tables of quantitative or qualitative measures indicating material being obtained, difficulty of obtaining the material and likelihood of detection. These results can be presented in various graphical or tabular forms, depending upon the needs of the individual user and the audience they will be presenting to.

6. Who would use the results?

   The range of users of the methodology includes designers, program policy makers, national regulators, international agencies, and other stakeholders.

7. What is the expertise needed to perform an evaluation?

   Familiarity with the PR&PP methodology, the system design, and the general requirements of non-proliferation (e.g., international safeguards) and physical protection is needed. Note that this combined expertise need not reside in a single evaluator, but can be represented by an assessment team.

8. How does a PR&PP evaluation differ from an IAEA INPRO evaluation of proliferation resistance?

   PR&PP is a design tool that evaluates the system response to a spectrum of potential PR&PP challenges (e.g., diversion, clandestine program, break-out scenario and terrorism), and can also be used by customers or policy makers in guiding decisions; INPRO is a best-practices checklist aimed mainly at assisting embarking countries in making decisions. INPRO is a broad assessment, including high-level state policy, while PR&PP is focused on the technology. [The International Project on Innovative Nuclear Reactors and Fuel Cycles (INPRO) was established by the International Atomic Energy Agency (IAEA) and has developed a methodology for proliferation resistance evaluation which can be found at http://www.iaea.org/INPRO/about.html.]

9. What is the relation between PR&PP and IAEA safeguards?

   PR&PP incorporates International Atomic Energy Agency (IAEA) safeguards as external components (sometimes referred to as "extrinsic measures") of the nuclear energy system being assessed. PR&PP can be used by designers at an early stage to assess where and how one might implement safeguards, in order to guide conceptual design decisions. Safeguards is one feature of a good design that enables the detection of proliferation. PR&PP assessment assures that this detection is timely by accounting for technical difficulty, time to achieve success, and material type.

10. Does PR&PP apply to national or sub-national actors?

    Generally PR applies to national actors, and PP to sub-national actors, although there can be interaction between these two.

11. How does a PR&PP evaluation relate to a safety and reliability evaluation?

    There are possible synergies between PP&PP and safety/reliability if assessed early enough in the design process and possible conflicts that may be addressed at this time. Even at the earliest stages of conceptual design, it is recommended that, in addition to representative safety challenges, designers perform qualitative assessments of the potential system response to representative PR and PP challenges.

12. At what stage of an evaluation should a PR&PP evaluation be performed?

    PR&PP can be performed at the earliest stages of design, and then revisited periodically and detail added as a design progresses and more detail is known.

13. To what type of systems does a PR&PP evaluation apply?

    PR&PP is designed for the assessment of nuclear energy systems, with no specific technology dependence. The elements of the fuel cycle (i.e., components of the fuel supply, reactor, or spent-fuel management architecture) that are included in the assessment are dependent upon the user's needs.

14. Are there other methodologies for performing similar evaluations?

    There are other assessment methodologies (e.g., INPRO). The methods are complementary. INPRO assessments can assure that best practices have been considered, and adopted where appropriate, in system design. PR&PP assessments can assure that the system response to PRPP challenges will be acceptable.

15. Will the PR&PP methodology be updated or is it in final form?

    The methodology is ready for use, currently as Revision 6. It will receive updates as feedback is received on implementation, and as the field of knowledge in non-proliferation and physical protection evolves.

16. What are benefits to performing PR&PP analysis early in the design process?

    Benefits include synergies with safety and economics, reduction of risk for schedule slippage, efficiencies in implementing PR&PP-related system design components, guidance on design decisions with maximum flexibility to early design stage.

17. Where can I find more information?

    The PR&PP Evaluation Methodology report [1] is available online. The journal Nuclear Technology provides several articles discussing the methodology and providing examples of its application [4].

5.3 Insight from Applications

5.3.1 *The Case Study*

The PR&PP methodology was developed and tested with the help of an example design. This was the Example Sodium Fast Reactor (ESFR). The design is described in detail in Reference [5]. Basic lessons learned from this case study included the following:

- Each PR&PP evaluation should start with a qualitative analysis allowing scoping of the assumed threats and identification of targets, system elements, etc.
- Detailed guidance for qualitative analyses should be included in the methodology
- Access to proper technical expertise on the system design as well as on safeguards and physical protection measures is essential for a PR&PP evaluation
- The use of formal expert elicitation techniques can ensure accountability and traceability of the results and consistency in the analysis.
- Qualitative analysis offers valuable results, even at the preliminary design level.
- Greater standardization of the methodology and its use is needed.

5.3.2 *Insights from Interaction with GIF SSCs*

A study was performed jointly by the PRPPWG and the six GIF System Steering Committees (SSCs). The report [6] from this study presents the status of proliferation resistance and physical protection (PR&PP) characteristics for each of the six nuclear energy systems selected by the Generation IV International Forum (GIF) for further research and development.

The interaction between the PR&PP Working Group and the GIF System Steering Committees (SSCs) has provided insights on the type of reactor system information that is necessary and useful to collect before one begins a PR&PP evaluation. For a given NES design, information that is particularly important to PR&PP will include potential fuel types (including high-level characteristics of fresh and spent fuel), fuel storage and transport methods, safety approach and associated vital equipment (for confinement of radioactivity and other hazards, reactivity control, decay heat removal, and exclusion of external events), and approach to physical arrangement as it affects access control and material accounting for fuel (a potential theft target) and access control to vital equipment (a potential sabotage target).

5.3.3 *Study of Four Reactor Designs*

A multi-laboratory team of U.S. subject matter experts, including past and present members of the PR&PP Working Group, used the PR&PP evaluation methodology as the basis for a technical evaluation of the comparative proliferation potential associated with four generic reactor types in a variety of fuel-cycle implementations. These are a sodium fast reactor, a high temperature gas reactor, a heavy water reactor, and a light water reactor. The evaluation team undertook a systematic assessment, capturing critical assumptions and identifying inherent uncertainties in the analysis. A summary of the study was presented at the Institute of Nuclear Materials Management (INMM) 51st Annual Meeting [7].

5.3.4 *Further Insights*

The relevance of the insights varies based on the various stakeholders of a PR&PP evaluation: policy makers, system designers, and the safeguards and physical protection communities.

For policy makers:

- An assessment of the proliferation potential of a particular reactor design in nuclear energy system should consider the system's overall architecture, accounting for the availability and flow of nuclear material in the front and back end of the fuel cycle.

For designers:

- Designers can incorporate features in the design to facilitate easier, more efficient and effective safeguards for inspection and monitoring. For example, minimizing the number of entry and exit points for fuel transfer between system elements will enhance material containment, protection and accountancy (MCP&A), thus partially compensating for any lack of knowledge continuity by visual inspection during a fuel transfer.
- Material type for proliferation resistance is related to the chosen composition of the nuclear material. The designer can optimize the design either to reduce the material's attractiveness (e.g., increase burnup in the uranium fuel to raise the fraction of Pu-238, thereby lowering the quality of plutonium in the spent fuel), or to make post-acquisition processing of the material more complex, indirectly increasing the technical difficulty for the proliferator.

For safeguards inspectors:

- Augmenting inspections for handling and storing fresh and spent fuel would reduce proliferation potential.
- Enhanced inspection of fresh fuel would reduce the proliferation potential of covert diversion and misuse.
- Optimizing material type and material movement pathways to facilitate accountability measurements can make verification more effective and efficient.

5.4 Summary

The PR&PP methodology provides a framework to answer a wide variety of nonproliferation and security-related questions for NES and to optimize these systems to enhance their ability to withstand the threats of proliferation, theft, and sabotage. The PR&PP methodology provides the tools to assess NES with respect to the nonproliferation and security.

PR&PP analysis is intended to be performed, at least at a qualitative level, from the earliest stages of system design, at the level where initial flow diagrams and physical arrangement drawings are developed, and simultaneously with initial hazards identification and safety analysis. The reader is encouraged to see the collection of journal articles on PR&PP that is contained in a special issue of the American Nuclear Society's Nuclear Technology [4]. This journal issue contains numerous articles on methods and applications of PR&PP and related approaches.

While the methodology has been explicitly developed for advanced systems, it also has application to current nuclear systems.

Acknowledgements The author is grateful to the many members of the PRPPWG (past and present) in the representing GIF countries and organizations who have contributed to the methodology and promoted its applications over the past decade.

References

1. PR&PP Methodology, Revision 15 Sept 2011: https://www.gen-4.org/gif/jcms/c_40413/evaluation-methodology-for-proliferation-resistance-and-physical-protection-of-generation-iv-nuclear-energy-systems-rev-6
2. IAEA State Level Approach 2010, See, for example, http://www.iaea.org/safeguards/DDG-Corner/dg-statements-repository/TheFutureofSafeguards.html
3. A good discussion of physical security of novel reactor systems can be found in the NEI Position paper on physical security for small modular reactors, 31 July 2012; See http://pbadupws.nrc.gov/docs/ML1222/ML12221A197.pdf
4. Special Issue of the American Nuclear Society's journal Nuclear Technology (vol 179, Number 1, 2012)
5. PR&PP case study report, 15 July 2011; https://www.gen-4.org/gif/jcms/c_40415/esfr-case-study-report
6. Proliferation resistance and physical protection of the six generation IV nuclear energy systems, 15 July 2011; https://www.gen-4.org/gif/jcms/c_40414/proliferation-resistance-and-physical-protection-of-the-six-generation-iv-nuclear-energy-systems
7. Zentner M, Therios I, Bari R, Cheng L, Yue M, Wigeland R, Hassberger J, Boyer B, Pilat J, Rochau G, Cleary V (2010) An expert elicitation based study of the proliferation resistance of a suite of nuclear power plants. In: Proceedings of the 51th Institute of Nuclear Materials Management Annual Meeting, Baltimore, 11–15 July 2010

Chapter 6
Next Generation Nuclear Security Policy: Education, Research, and Experience

Erika Suzuki, Bethany Goldblum, Robert Brown, Stanley Prussin, and Michael Nacht

Abstract The Nuclear Science and Security Consortium (NSSC) is a multi-institution initiative composed of seven universities and four national laboratory partners working collectively to train the next generation of nuclear security experts. The NSSC draws students and scholars together in unconventional ways, replacing the boundaries that separate disciplines with a more inclusive science-technology-policy interface. To address the need for knowledgeable practitioners in the nonproliferation field, education and training in nuclear security policy is accomplished via three major activities: First, *Nuclear Security*: *The Nexus Between Policy and Technology* is an innovative classroom experience offered jointly by the Department of Nuclear Engineering and the Goldman School of Public Policy to educate students on the policy and technological foundations of nuclear energy and weapons. Second, the Nuclear Policy Working Group was established to provide students beyond-the-classroom opportunities to collaboratively engage in interdisciplinary nuclear security policy research projects anchored in science and technology. Third, the NSSC collaborates with the Institute on Global Conflict and Cooperation (IGCC) to provide immersive education in nuclear security policy. Since 2011, NSSC students and faculty have participated in the Public Policy and Nuclear Threats Boot Camp, an established multidisciplinary program offered by the NSSC and the IGCC designed to provide graduate students and mid-career professionals with the necessary tools to contribute to the debate on nuclear security policy.

E. Suzuki • B. Goldblum (✉) • S. Prussin
Department of Nuclear Engineering, University of California, Berkeley, USA
e-mail: bethany@nuc.berkeley.edu

R. Brown
Department of Political Science, Temple University, Philadelphia, PA, USA

M. Nacht
Goldman School of Public Policy, University of California, Berkeley, USA

S. Apikyan, D. Diamond (eds.), *Nuclear Threats and Security Challenges*,
NATO Science for Peace and Security Series B: Physics and Biophysics,
DOI 10.1007/978-94-017-9894-5_6

6.1 Overview

In 2011, the National Nuclear Security Administration (NNSA) awarded the Nuclear Science and Security Consortium (NSSC) $25 M to establish a 5-year program seated at the University of California (UC), Berkeley, to train the next generation of nuclear science and policy leaders [1, 2]. The NSSC's mission is to support the nation's nuclear security agenda by recruiting students of the highest caliber to study the core principles of nuclear science in preparation for research and leadership roles in the NNSA architecture. The consortium is a network of universities and national laboratories; composed of seven academic institutions, including UC Berkeley, UC Davis, UC Irvine, the University of Nevada, Las Vegas (UNLV), Michigan State University (MSU), Washington University at St. Louis (WUSTL), and the Institute on Global Conflict and Cooperation (IGCC) at UC San Diego; and four national laboratories, including Lawrence Berkeley, Lawrence Livermore, Los Alamos, and Sandia National Laboratories.

The NSSC policy focus area provides academic coursework, research opportunities, and hands-on experience and immersion in nuclear security policy built on a foundation of nuclear science. Whereas other consortia or initiatives, such as the Next Generation Safeguards Initiative [3] and the NNSA's Global Threat Reduction Initiative [4] have heavily emphasized either graduate student research and training or the development of rigorous academic coursework, the NSSC has designed a nuclear security policy program that combines innovative coursework, extracurricular research opportunities, and hands-on experience in national laboratories for students at the undergraduate, graduate and postdoctoral level to develop a well-rounded, and capable cadre of nuclear security professionals. This is accomplished through a multi-pronged approach that includes an interdisciplinary course on nuclear security, research opportunities through the Nuclear Policy Working Group, and an intensive 10-day training session via the Public Policy and Nuclear Threats Boot Camp.

6.2 Education for the Next Generation

In developing the *Nuclear Security* course, the NSSC sought to address an omission in the curriculum of many academic institutions and fill a gap in the course offerings at UC Berkeley. The NSSC *Nuclear Security* course (NE/MPP 285) is the only nuclear-security-focused course on campus, and is one of only a handful of multidisciplinary courses that bridge policy and science. As UC Berkeley is the leading organization for the nuclear security policy focus area, the course was developed to pioneer next generation education in nuclear security science and policy, and serves as a flagship course for the development of similar courses at partner NSSC academic institutions. Michael Nacht, Professor at the Goldman School of Public Policy and NSSC Nuclear Security Policy focus area lead, provided policy expertise developed over decades of experience in the United States (US)

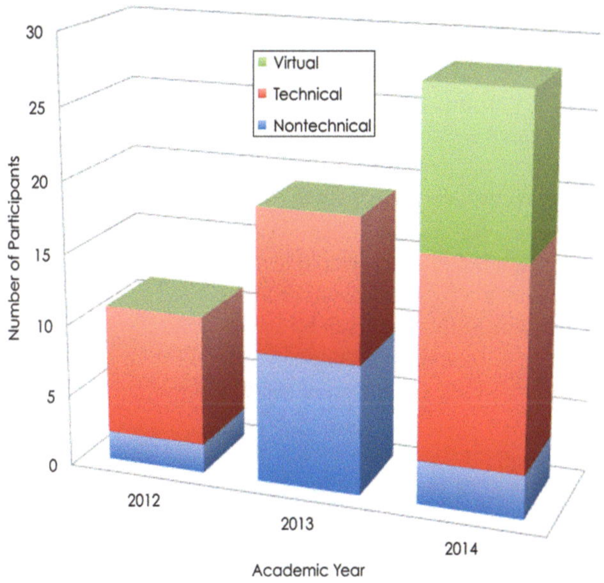

Fig. 6.1 Demographics for the nuclear security: the nexus between policy and technology course. The number of technical, nontechnical, and "virtual" participants in the course is listed per academic year

government and national security infrastructure, while Stan Prussin, Professor in the Graduate School for the Department of Nuclear Engineering at UC Berkeley, provided technical expertise in fundamental and applied nuclear science, acquired as a result of decades of cutting-edge research in applied nuclear physics. Together, the instructors developed a multidisciplinary curriculum to provide students from a variety of backgrounds and educational stages with the knowledge and tools required to contribute to nuclear security policy. First offered in Spring 2012, the course is delivered annually through the Goldman School of Public Policy and the Department of Nuclear Engineering at UC Berkeley, as well as via a distance-learning platform. As shown in Fig. 6.1, the number of course participants continues to rise. Over a 3 year period, NE/MPP 285 students have represented over 15 different majors in both the technical and social scientific fields, ranging from nuclear engineering and physics to political science and philosophy. Though the course is offered at the graduate-level, the instructors have made case-by-case exceptions for interested undergraduate students to enroll in the course, noting the importance of early-stage education and exposure to nuclear security issues. As a result, undergraduate students ranging from freshmen to seniors account for one-third of the total course participants.

The course curriculum addresses historical and current events, and features in-depth case studies that highlight today's most pressing security concerns. Lecture topics vary by year as the instructors reach to address current security issues, such as the North Korean nuclear threat, the Iranian nuclear program, and interdicted nuclear material. The instructors employ a dual-track teaching method to minimize knowledge gaps between the social scientists and technical students. For the first half of each course meeting, social scientists and humanities students are trained in fundamental nuclear physics and scientific applications in radiation detection, nuclear reactors and weapons design, and more. Contemporaneously, students from technical science and engineering fields are taught the fundamentals of international relations, government, and public policy. The second half of the course meeting brings all students together in one room to hear a joint-lecture on a topic in nuclear security, drawing on concepts and lessons from both sessions in the first half of the course meeting. By separating students by background, the instructors are able to minimize knowledge gaps and level the playing field for all students, who can then engage with peers from radically different fields and backgrounds on the subject of nuclear security.

The students are given opportunities to practically apply material covered in lecture and discussion through a mid-semester crisis simulation and a semester-long research project. The crisis simulation, conducted in two phases, enables students to sit in the seats of powerful decision-makers in the US government, such as the President or the Chairman of the Joint Chiefs of Staff, and work collaboratively under time pressure to address a mock crisis situation. For example, the Spring 2014 crisis simulation provided the following scenario:

> It is late May 2014. The situation in Ukraine has worsened substantially, and parts of eastern Ukraine are now in civil conflict between pro-Ukrainian and pro-Russian forces. President Obama has imposed sectoral sanctions against the Russian oil and natural gas, banking, and technology sectors. In response, President Putin has stipulated that Russia is advising Iran not to follow through on the terms of the interim agreement. A special P5+1 negotiating session with Iran is scheduled for late May to determine whether the interim agreement can in fact be converted into a permanent treaty that would halt or roll back Iran's nuclear weapons program.
>
> —Spring 2014

The first phase of the simulation exercise was a US-Israel negotiation on how to approach the Iran nuclear discussions; the second phase involved a US-Iran negotiation simulating the negotiations currently in progress [5]. The students were assigned roles as US government representatives, Israeli government representatives, and Iranian government officials. Following the US-Israel discussions that highlighted major differences between the two governments, the US-Iran negotiations were more extensive and led to possible compromises in future rounds. Students delivered options for negotiating a deal with Iran, as well as formulating a response to Russia in the wake of the Ukraine conflict. The students had to work under time pressure and overcome communication challenges to fully understand available policy options informed by technical information about Iran's nuclear program.

The semester-long collaborative research project provides students with another opportunity to practically apply lessons learned, and challenges students to overcome communication barriers across the technical and nontechnical fields in order to develop an informed policy recommendation built on a clear understanding of the nuclear science and technological issues. Hypothetical project topics are also drawn from current events, and all enrolled students must work together to coordinate a coherent response. For example, the Spring 2014 topic tasked students with preparing a paper outlining how the US should proceed in the event of a nuclear attack on Moscow by Chechen rebels. The instructors required students to consider available technical and intelligence information, and technical pathways that could be pursued to fill information gaps and help establish that the US was not involved. Though the instructors provided guidance and support throughout the semester, students had to work collaboratively and independently to overcome communication challenges across academic cultures and effectively organize into sub-topic groups to explore specific components of the project. The project stimulates close interaction across diverse disciplines and also promotes the development of valuable professional skills in project management, communication, and leadership.

The *Nuclear Security* course also features a distance-learning component, which enables students from other NSSC academic institutions to engage in the course. Initially, course lectures were available as recordings and distributed to affiliated NSSC faculty and students. In Spring 2014, the course was made broadly available through a live-streaming webcast. As a result, 11 virtual auditors from NSSC partner institutions participated in the course in real time (see Fig. 6.1). Through the distance-learning and archiving platform, the NSSC has engaged a broader cross section of students at partner academic institutions while ensuring that course lessons will be available for years to come.

Retention of social scientists and humanities students remains a challenge for the course. The instructors have noted that technical jargon, equations, and graphs intimidate nontechnical students who are not confident about overcoming a steep learning curve to understand fundamental nuclear science. In order to recruit and retain more nontechnical students, a *Nuclear Science for Future Policymakers* workshop was fielded in January 2015. The workshop focused on the relationship between science and policy, and featured lectures on fundamentals of radiation and radioactivity, the nuclear fuel cycle, and nuclear weapons design. A radiation detection practical activity provided nontechnical students with valuable hands-on experience.

6.3 Nuclear Security Policy Research

In contrast to research projects facilitated under formalized programs such as the Undergraduate Research Apprenticeship Program [6] or other narrowly defined research assistantships on campus, the Nuclear Policy Working Group (NPWG) provides a unique forum in which students develop their own interdisciplinary research projects in nuclear security policy [7]. The NSSC Director of Education, Dr. Bethany Goldblum, founded the NPWG in Fall 2012 to educate students on important issues in nuclear security, foster close and continuing collaboration between students from diverse academic disciplines, and guide the development of student-led research projects and policy recommendations to contribute to the field of nuclear security. This low-cost, high-impact working group model is driven by student passion; the majority of the two-dozen active members of the NPWG are not paid and do not receive course credit for their participation in weekly seminars and collaborative research. The working group provides an environment in which undergraduate students can interact with and engage in collaborative research projects with graduate students, postdoctoral scholars, national laboratory scientists and researchers from a variety of nontechnical and technical fields.

The NPWG holds weekly interactive seminars, during which students explore topics in nuclear security policy, such as nuclear forensics, arms control, societal verification, and network science for nonproliferation. Guest speakers from the national laboratories, think tanks, other academic institutions and campus research centers provide expert guidance via lecture and discussion. The working group forum encourages freethinking and creative solutions to today's most challenging issues in nuclear security. In many high-level academic and national laboratory settings, students may be intimidated by the expertise of their research groups and potential criticisms from their peers to ask questions. The NPWG creates an environment in which there is no unreasonable question, and such a relaxed, supportive atmosphere enables students to explore basic questions, such as "*What is fission*?" and "*What is the Non-Proliferation Treaty*?" To further understanding of more complex concepts, such as radiation detection and international security policy, the NPWG also offers introductory workshops to supplement expert lecture and discussion.

Throughout the semester, students are encouraged to explore topics of their own interest in nuclear security through collaborative research projects. Research projects are largely open-ended; they are required to address a relevant issue in nuclear security and nonproliferation, and the topic must bridge nuclear security policy with one or more of the NSSC's technical focus areas: nuclear engineering, nuclear physics, radiochemistry, and radiation detection and instrumentation. As is common in the interdisciplinary field of nuclear security, projects often overlap with multiple technical focus areas. Multidisciplinary, student-led, research projects help students build crucial skills through hands-on experience in communication, problem-solving, and critical thinking. The NPWG research teams are commonly led by a graduate student and are composed of undergraduates, graduate students, and professionals from both nontechnical and technical fields. Team leads develop crucial skills in project management, leadership, and mentorship-skills highlighted as desirable traits by partner national laboratory program managers. Teams strive to craft coherent, informed recommendations targeting publication in an academic journal. Since the NPWG's inception, the group has published two papers and has several more under development [8, 9].

The NPWG not only provides unique research opportunities to undergraduates, the group also enables graduate students from the NSSC to explore the policy implications of their technical research. NPWG members and team leads at UC

Fig. 6.2 Dr. Raluca Scarlat (at *left*), a postdoctoral scholar with expertise in advanced nuclear energy system moderates the 2014 Expert Panel Discussion on April 29, 2014. Panelists include Dr. Zachary Davis, Senior Fellow at the Center for Global Security Research at Lawrence Livermore National Laboratory; Dr. Charles Ferguson, President of the Federation of American Scientists; Dr. Sabine Roeser, Professor of Ethics and Antoni van Leeuwenhoek Chair in the Department of Philosophy at TU Delft, Netherlands; and Dr. Per Peterson, Professor of Nuclear Engineering at UC Berkeley and former William and Jean McCallum Floyd Endowed Chair and a Commissioner for the Blue Ribbon Commission on America's Nuclear Future

Berkeley, Alexandra Asghari and Eva Uribe, are two examples of technical graduate students who are conducting complementary policy research to include in their doctoral dissertations. In addition, NSSC Nuclear Security Policy Fellow, Kalee Hammerton from MSU has worked with the NPWG at UC Berkeley to start a satellite NPWG chapter at MSU and explore the policy implications of her research in radiochemistry. Such activities connect scientific research and technology applications within a policy context. NPWG team members also gain valuable experience in presentation, networking, and communication by presenting their team's research at conferences and workshops.

The working group also engages with the UC Berkeley campus community and students at other NSSC academic institutions through a variety of educational outreach activities. Most prominently, the NPWG organizes an annual *Expert Panel Discussion on Nuclear Security Policy*. The purpose of the event is to raise awareness of important topics in nuclear security. In 2013, the theme of the panel event was *The Role of Nuclear Forensics in Nuclear Security Policy* [10]. As shown in Fig. 6.2, the 2014 event, focused on *Ethics in Nuclear Science and Security Policy* [11], featured lively discussion amongst a broad range of experts. Panelists drew from their diverse backgrounds in academia, the national laboratories, and think tanks to share their perspectives on topics related to the theme. For the 2014 event, the NPWG featured a live-streaming webcast to broadly engage students, faculty, and national laboratory scientists. Through interactive seminars, student-led investigations and such educational outreach, the NPWG facilitates the second pillar of nuclear security policy education, research, and training in the NSSC's policy program.

6.4 A Legacy in Nuclear Security

The Public Policy and Nuclear Threats Boot Camp (PPNT) is a decade-old workshop-in-residence developed by IGCC and held at UC San Diego [12]. The boot camp was originally designed as a graduate-level training program funded by the National Science Foundation's Integrative Graduate Education and Research Training program. In 2011, the NSSC partnered with IGCC to complement its technical training with immersive education in nuclear security policy. The 2014 PPNT Boot Camp marked a shift in leadership, as NSSC Director of Education and PPNT alumna, Dr. Bethany Goldblum, served as Program Director, furthering the legacy of former Program Director and fellow PPNT alumnus, Dr. Robert Brown. Ambassador Linton Brooks, who has served as a senior expert and scholar-in-residence at PPNT since 2008, continued in this capacity, sharing decades of experience in arms control and diplomacy and facilitating lively discussion and engagement among participants and speakers (see Fig. 6.3).

The boot camp agenda provides a comprehensive and immersive curriculum spanning issues in nuclear security, including nonproliferation, safeguards, arms control, nuclear forensics, terrorism, radiation detection and interdiction, and more. PPNT participants hear from government officials, national laboratory scientists, public policy practitioners, and representatives from international organizations, such as the IAEA, on past and ongoing nuclear security efforts. Participants also have the opportunity to engage with expert speakers outside the classroom in a candid and open environment that fosters discussion and debate about how to address today's most pressing concerns. Expert speakers become mentors to participants, imparting career advice and providing insight into a variety of pathways to become effective nuclear security professionals.

In addition to engaging with today's leaders in nuclear security, PPNT program participants network with one another and practice communicating diverse perspectives across different disciplines. Relationships developed during the program are often lasting, as evidenced by a strong network of over 200 past PPNT participants. Hands-on sessions like the radiation detection practical led by Dr. Mark Schanfein from Idaho National Laboratory enabled nontechnical students to handle detectors and learn about role of such equipment in the larger nuclear security policy architecture. NSSC technical students from the radiation detection and instrumentation focus area worked with Dr. Schanfein to show nontechnical participants and students from other technical fields how to operate the portable detectors and

Fig. 6.3 Dr. Bethany Goldblum, Amb. Linton Brooks, and Dr. Robert Brown kick-off the 2014 PPNT Boot Camp

scan commercially available sources. An extended 2-day session designed by Prof. Michael Nacht and Dr. Joseph Pilat from Los Alamos National Laboratory, and facilitated by Dr. Pilat and Ambassador Brooks was added to the program this year. On the final day of the program, participants put their knowledge and skills to the test in a simulation session facilitated by Ambassador Brooks. Participants were split into three groups and responded to nuclear crisis scenarios under time pressure and with limited information (see Fig. 6.4). Like the simulation in the nuclear security course, the PPNT simulations challenge participants to overcome communication barriers, information gaps, coordination difficulties, and opposing opinions to develop a coherent and targeted response. Each group had to present their policy briefs to a panel of "government officials" composed of Program Director, Dr. Bethany Goldblum, Ambassador Linton Brooks, and NSSC Advisory Board Chair and President of the Hertz Foundation, Dr. Jay Davis.

6.5 NSSC and the Future of Nuclear Security

The NSSC strategy is necessary for ensuring that the future nuclear security workforce is prepared to continue the legacy of current practitioners. Targeted research must complement academic work, and immersion programs that provide

Fig. 6.4 Participants of the 2014 PPNT Boot Camp present their policy brief in a simulation exercise

hands-on experiences and networking opportunities for budding professionals is crucial. Academic coursework is not enough; students also need immersion in the field and hands-on experience to develop communication, project management, and leadership skills. NSSC students are encouraged to participate in all three programs to develop the knowledge base and skills necessary to become leaders in nuclear security science and policy.

Activities are currently under way to preserve lessons in nuclear science and security and provide more opportunities for engagement in policy research and education. Profs. Nacht and Prussin are working on a textbook based on course lectures to capture lessons and perspectives in nuclear science, policy, and technology, and pass on best practices and insight to the next generation. To provide similar opportunities for extracurricular research, the NPWG is expanding to other NSSC consortium schools and affiliated Minority Serving Institutions to establish satellite working groups through its Nuclear Security Education Initiative. The group will also continue to develop educational programming, professional development, and outreach activities to prepare the next generation to address threats to nuclear security. The NSSC Director of Education, Dr. Bethany Goldblum, will again lead the PPNT program next year, and welcome a new cadre of technical NSSC students, social science students and professionals. Via this multifaceted nuclear security program, the NSSC is training the next generation of nuclear science, technology and policy leaders [13].

Acknowledgements This material is based upon work supported by the Department of Energy National Nuclear Security Administration under Award Number. DE-NA0000979. This report was prepared as an account of work sponsored by an agency of the United States Government. Neither the United States Government nor any agency thereof, nor any of their employees, makes any warranty, express or limited, or assumes any legal liability or responsibility for the accuracy, completeness, or usefulness of any information, apparatus, product, or process disclosed, or represents that its use would not infringe privately owned rights. Reference herein to any specific commercial product, process, or service by trade name, trademark, manufacturer, or otherwise does not necessarily constitute or imply its endorsement, recommendation, or favoring by the United States Government or any agency thereof. The views and opinions of authors expressed herein do not necessarily state or reflect those of the United States Government or any agency thereof.

References

1. Nuclear Science and Security Consortium, University of California, Berkeley (2014) www.nssc.berkeley.edu
2. Goldblum BL, Vujic, J (2014) Nuclear Science and Security Consortium: training the next generation. In: Proceedings of the Institute on Nuclear Materials Management 55th annual meeting. Atlanta
3. Next Generation Safeguards Initiative, National Nuclear Security Administration (2009) http://nnsa.energy.gov/mediaroom/factsheets/nextgenerationsafeguards
4. Marianno CM, Charlton WS, Contreras AR, Unlu K, Lanza RC, Kohse GE (1999) Designing the global threat reduction initiative's nuclear security education programs. University of Texas, A&M, College Station
5. Iranian, US representatives hold 6 hours of nuclear talks, Iran Republic News Agency, September 19, 2014
6. Undergraduate Research Apprentice Program, University of California, Berkeley (2014) http://research.berkeley.edu/urap
7. Nuclear Policy Working Group, Nuclear Science and Security Consortium, University of California, Berkeley (2014) www.npwg.berkeley.edu
8. Goldblum, BL, Suzuki E, Vujic J (2013) Building a foundation for the future of nuclear security. J Sci Secur Publ Interest Rep, Federation of American Scientists 66(3)
9. Suzuki E, Tregub I (2014) Would Japan be worse off building the bomb? An analysis of international and domestic security considerations. Nuclear notes, Center for Strategic and International Studies, project on nuclear issues 4(1):50–57
10. The Role of Nuclear Forensics in Nuclear Security Policy, Nuclear Policy Working Group, University of California, Berkeley (2013) http://npwg.berkeley.edu/events/2013panel
11. Ethics in Nuclear Security Science and Policy, Nuclear Policy Working Group, University of California, Berkeley (2014) http://npwg.berkeley.edu/events/2014panel
12. Public Policy and Nuclear Threats Boot Camp, Institute on Global Conflict and Cooperation, University of California, San Diego (2014) http://www-igcc.ucsd.edu/workshops/public-policy-and-nuclear-threats
13. Suzuki E, Goldblum, BL, Nacht M, Prussin S, Brown RL, Vujic J (2014) Nuclear Security Science and Society: connecting students to scientists. In: Proceedings of the Institute on Nuclear Materials Management 55th annual meeting, Atlanta

Chapter 7
The Management of Radioactive Waste in France: What Is the Threat?

Jean-Michel Boniface and Gérald Ouzounian

Abstract Radioactive waste is a potentially valuable material for a terrorist organization that could use it against the population with malicious intent. This article addresses the key aspects of this issue through the prism the radioactive waste management, and illustrates how this security challenge has been considered in the French electro-nuclear field. Indeed, with 58 nuclear power reactors each producing yearly a significant amount of various types of radioactive materials, the management of the final waste product is of paramount importance in regard to the long-term safety of the population and the protection of the environment, as well as in terms of security threat mitigation.

Concerning the LILW-SL radioactive waste, which also might be suitable for a potential terrorist plan, the final conditioning of the materials stabilizes the waste form and prevents all potential release of contaminated particles inside the dedicated containers. Such methodology provides a solid barrier against any unauthorized attempt to retrieve the active substance. The disposal facilities developed in France by Andra, have been designed and are operated with the view to mitigate the risk of pollution and to limit the risk of voluntary intrusion. The concept of a near surface repository that includes engineered barriers consisting of concrete vaults drastically decreases the possibility of attacks with the objective to extract the radioactive inventory disposed of on the site.

The proliferation threat is much more relevant when it comes to the spent nuclear fuel. It should be considered with all due attention during all stages of the spent fuel management. France re-cycles the spent fuel and the final residues are immobilized within a glass matrix and packed into canisters that make the material much less "attractive" for proliferation purposes. Considering the French project of a deep geological repository for HL radioactive waste, also developed by Andra, the waste package retrievability is one of the major design-driving characteristics. A possibility of malicious attempts to retrieve the waste for proliferation is therefore

J.-M. Boniface (✉) • G. Ouzounian
ANDRA, Châtenay-Malabry, France
e-mail: Jean-michel.BONIFACE@andra.fr

S. Apikyan, D. Diamond (eds.), *Nuclear Threats and Security Challenges*,
NATO Science for Peace and Security Series B: Physics and Biophysics,
DOI 10.1007/978-94-017-9894-5_7

considered a risk and is taken into account already at the design phase. This experience of Andra is also taken into account in the frame of IAEA's ASTOR group work.

As with infamous nuclear disasters of 1986 and 2011, contaminated materials could also become an easy target for a terrorist organization. In this respect, Andra has developed a quick-to-implement disposal solution designed to accommodate large amounts of low-level radioactive materials. This solution ensures the ultimate safety for the surrounding population and the environment, and reduces the possibilities of repository violation.

In response to the proliferation threat and other potential security concerns, France has developed an effective and comprehensive radioactive waste management plan that prioritizes the safety of the population and the protection of the environment, and provides a high level of security in this domain.

7.1 Introduction

Since time immemorial, both sword and shield have been an absolute necessity in order to protect populations from threats orchestrated by organisations capable of ingeniously exploiting the technical, technological and social changes of our civilisations. There are numerous examples to illustrate how the slightest chink in the armour has enabled some to transform everyday objects or situations into veritable killing machines. Air transport has certainly been the field in which this has been illustrated with the greatest violence and cruelty in recent decades. More insidiously, large scale information exchanges, spectacularly amplified by the internet, have enabled small, ill-intentioned groups to weave their terrorist webs across the planet. Recently, the threat to Haditha dam in Syria was felt to be serious enough to warrant air strikes, in order to prevent this electricity generating facility from being transformed into a weapon of mass destruction. Unfortunately, there could be further examples.

What about nuclear energy? Its power plants, its facilities, its uranium, its research centres? These are all aspects of vital importance which could become the prey or the target of a malicious act. Of these, one in particular is the focus of attention by Andra, the French national radioactive waste management agency: how ultimate radioactive materials, considered by the nuclear power generating industry to be waste, with no possible reuse nor recycling value, could nonetheless appear to radical groups as real tools of terror capable of threatening entire regions. Because the goal of these groups is to spread terror by all physical or psychological means, to force the World's leaders to comply with their wishes, it is possible to imagine that simply evoking the possible dissemination of radioactive material at strategic points of a country, regardless of its level of radiological activity, would be enough to create panic and seriously disrupt the working of a State. The Sarin gas terrorist attack in Japan in 1995 is still fresh in the memory.

However, radioactive waste is today certainly not the first target for acquisition by terrorists, but we must not allow them too much time to consider the possibilities and must examine the existing means of protection and deterrence to prevent this threat from becoming reality.

This article proposes an overview of the situation in France, and how Andra helps to keep this threat down to an acceptable level.

7.2 Andra: The French National Radioactive Waste Management Agency

In 1991, the French government created the National Radioactive Waste Management Agency (Andra), whose task is to find, design and implement safe management solutions for all French nuclear waste. For this purpose and since its creation, Andra developed and has been operating for now more than 20 years the disposal facilities for French LLW/ILW, one in the Manche departement and the other in the Aube departement, 300 km from Paris. With a total capacity of more than 1.5 million m^3, these Andra disposal facilities receive radioactive waste in the form of metal drums or in concrete containers, which are then placed in vaults, where the disposal conditions will preserve the populations and environment from all radiological pollution for at least 300 years, by which time the residual activity of the materials will have sufficiently decayed so that they no longer present a hazard.

For the past 15 years, Andra has been studying the conditions for deep geological disposal of the highly radioactive materials which come primarily from recycling of the spent fuels produced by the nuclear power plants.

Although for LLW/ILW waste the risk of proliferation would appear to be very slight, or even non-existent, the same cannot be said for the high-level materials. This is why, from the outset, Andra has been developing expertise not only in radiological protection tailored to the waste disposal conditions, but also in the field of engineering, for the physical protection of its facilities and the materials disposed of therein. Therefore the reports prepared by the scientists and engineers at Andra meet the requirements set by the French nuclear safety regulator (ASN) and the security constraints designed to minimise the risk of intrusion and malicious acts.

7.3 Radioactive Waste in France

With 58 nuclear power reactors producing power to meet 75 % of the country's electricity needs, France has for many years been developing an ambitious programme to manage the radioactive waste produced by these plants, the related industries, notably the La Hague plant for recycling spent fuels, and also from research and other facilities working outside the nuclear power generating sectors (hospitals,

Table 7.1 Classification of radioactive materials in France (Source Andra)

French classification of radioactive waste and management solutions (existing or being considered)

	Half-life		
	Very short-lived	Short-lived	Long lived
Activity level	<100 days	<30 years	>30 years
Very low-level waste (VLLW)	Waste managed by allowing the radioactivity to decay in situ	Aube VLL waste disposal facility (CIRES)	
Low-level waste (LLW)		Aube LLW/ILW waste disposal facility (CSA)	Studies in progress (radium-bearing and graphite waste)
Intermediate-level waste (ILW)		Studies in progress (tritiated waste)	Studies in progress pursuant to the 1991 Act
High-level waste (HLW)		Studies in progress pursuant to the 1991 Act (CIGEO project)	

manufacture of specific equipment, etc.), each of which produced various quantities and types of materials to be processed and managed.

In 2006, the French Government established the National Plan for Radioactive Materials and Waste Management (PNGMDR), following extensive preparatory work initiated 10 years previously by ASN, and in which Andra was already a participant. The aim was to define fundamental rules for effective and acceptable short- and long-term management of radioactive elements.

To do this, the PNGMDR relies on an exhaustive inventory of the quantities of radioactive materials present in the country, regardless of producer, and gives an annual update of the future quantities according to the industrial scenarios envisaged and the options selected by the country's political authorities in order to define the energy mix for the coming decades. This inventory is drawn up and published by Andra every 3 years and is available on its website www.andra.fr.

The preparation of this inventory is above all based on the classification commonly adopted for radioactive materials and on the existing or future solutions for disposal of the waste. This classification and these solutions are presented in Table 7.1.

The inventory published in 2012 gives existing quantities and those estimated for the coming years. It is presented in Table 7.2.

The inventory is hence based on a wide-ranging dialogue between the waste producers, the public authorities and Andra, in order to define the volume of radioactive materials to be managed, according to the electrical energy production scenarios. The scenarios studied in France and on which the inventory forecasts are based, primarily concern the service life extension of the nuclear power plants from 40 to 50 years, the changes in the French nuclear power reactor fleet and the varying share of renewable energy sources in the country's energy mix.

Table 7.2 Inventory published in 2012 (Source Andra)

	2010 (m^3 conditioned waste)	2020	2030
HLW	2,700	4,000	5,300
ILW-LL	40,000	45,000	49,000
LLW-LL	87,000	89,000	133,000
LLW/ILW-SL	830,000	1,000,000	1,200,000
VLLW	360,000	762,000	1,300,000
Total	~1,320,000	~1,900,000	~2,700,000

Whatever the scenario finally adopted, the existing and future quantities will represent considerable volumes which demand the full attention of the stakeholders in the field, in order to minimise the risk of misappropriation of these materials, even if only a few cubic metres, to organise acts of terror.

7.4 LLW/ILW-SL Waste

As it contains no fissile products, low and intermediate level, short-lived waste presents no risk in terms of nuclear proliferation. The danger that could arise from the recovery of this material by an ill-intentioned undertaking could primarily lie in the psychosis of terror created by the announced dissemination of material that is contaminated or which emits radiological radiation. The paralysis of the nation's services and infrastructures could seriously destabilise a country and help bend it to the will of the perpetrators of these acts.

However, in France LLW/ILW waste is packaged at source by the producers in such a way that the most heavily ionising elements are immobilised in a solid matrix, generally made of concrete, and placed in metal drums. This matrix also entails significant dilution of the specific activity of the elements concerned, thus considerably reducing the radiological dose emitted at the surface of the packages. Once they reach the disposal site, Andra again places these drums in reinforced concrete containers and immobilises them with a sealing mortar. Ill-intentioned recovery of radioactive material is then impossible and, at worse, would only lead to pieces being chipped off, with an ability to harm that would obviously be far from that being sought.

7.5 High-Level Waste

France has opted for recycling of the spent fuels from the nuclear power plants. There are many reasons for this: production of reusable MOX, reduced volume of high level ultimate waste to be managed, etc. The volume of high level material produced annually in France is about 180 m^3. The chemical process utilised since the 1970s in the La Hague recycling plant separates the actinides and other reusable fission products of Uranium and Plutonium and ends with encapsulation of the high level ultimate waste ($>10^8$ Bq/g) in a glass matrix offering mechanical, chemical and containment stability, before being poured into stainless steel canisters.

The final disposal of these canisters is scheduled to start in 2028 in the CIGEO deep geological disposal facility. By then, the first waste packages, mainly stored on the Areva sites for more than 40 years, will have cooled sufficiently for them to be transferred to this underground environment.

In these conditions, any attempt to recover the fissile material trapped in this glass matrix, assuming that it were possible for the intruder to reach the canister disposal location, would require a gigantic industrial infrastructure and a logistical organisation that would be hard to hide during the preparation phase.

7.6 The Disposal Facilities in France

France currently possesses two disposal facilities for low and intermediate level short-lived waste (LLW/ILW-SL), one for the disposal of very low level (VLL) waste and is examining the disposal facility for future high-level or low and intermediate level, long-lived waste (LLW/ILW-LL).

The Manche facility (CSM) for LLW/ILW, created in 1969, received its last waste package in 1994 and entered the surveillance phase in 2003. It has a disposal capacity of 527,000 m^3. Its structure mainly consists of trenches topped by a multi-layer cover 4–6 m thick, comprising a succession of drainage and containment elements consisting of clay, sand and watertight geo-membranes. The waste is packaged in metal drums of 200 l which are then placed in the trenches, themselves filled with an immobilising matrix consisting of semi-draining material. In 1994, France opted to implement active surveillance of the CSM, not only to ensure a daily check on the effectiveness of the radiological barriers in place, but also to deter and prevent any attempted break-in.

The Aube Facility, commissioned in 1992 and of a more recent design than the CSM, has a total packaged waste disposal capacity of 1 million m^3, and will enter the surveillance phase in about 2050. The disposal units consist of reinforced concrete vaults, inside which the waste packaged in metal drums or concrete containers is emplaced and then mechanically immobilised with a concrete or gravel mortar. Once completely filled, the vaults are then closed by an 80 cm thick reinforced concrete cover, itself covered with a 4-m thick protective layer of clay and sand. This CSA design, now recognised as one of the worldwide benchmarks for the surface disposal of LLW/ILW waste, makes any intrusion virtually impossible.

With regard to very low level (VLL) waste, Andra commissioned a new facility (CIRES) in 2002, which will eventually be able to take 630,000 m^3 of waste. The design is based on disposal trenches, with the waste packaged in "big bags" or metal drums, before being covered by a clay layer of several metres, offering effective protection against malicious intrusion.

Finally, Andra has for several years been studying the future deep disposal facility for high level or long-lived waste. This future facility, scheduled to take the first packages in about 2028, will be situated at a depth of 500 m, in the middle of a homogeneous clay layer 150 m thick. The waste packages will be packaged in concrete containers and lowered into the disposal drifts via an access ramp, which will be the only means of reaching the disposal area. Once filled with waste packages, the drifts will be closed and sealed by a plug of bentonite clay, which has already clearly proven its strength. In these conditions, it would be hard to envisage the retrieval of radioactive material for terrorist purposes without those doing so exposing themselves to high radiological doses, such as to seriously compromise their macabre undertaking.

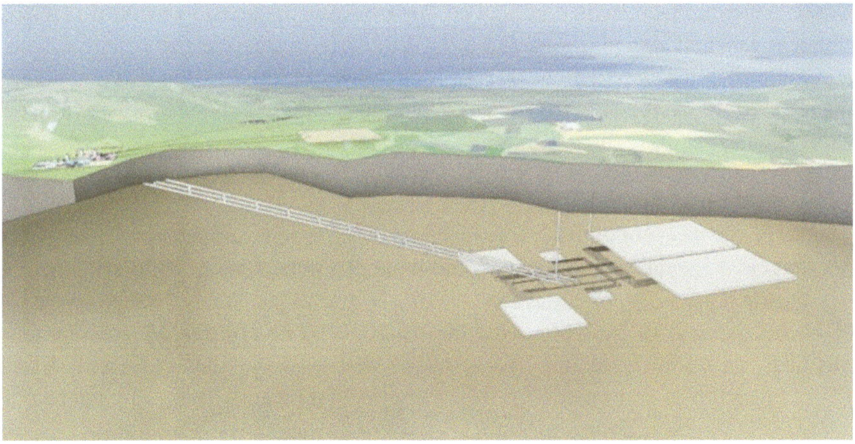

7.7 Waste Resulting from Post-accident Situations

The 2011 nuclear disaster was a timely reminder that there is no such thing as zero risk, despite all the precautions taken by Governments when developing their nuclear programmes. The human, economic and environmental disaster must not be compounded by a further threat linked to the possible recovery of thousands of cubic metres of rubble and soil contaminated by the accidental releases. Although the zones concerned are rapidly declared exclusion zones, it would be highly probable that ill-intentioned groups would be able to convince their followers to obtain soiled and potentially contaminated materials. As of 2005, well before the Fukushima Daiichi accident, France initiated the study of a Post-Accident Management Plan, run by ASN, within a Management Steering Committee (CODIRPA) to define the rules and procedures to be implemented in the initial aftermath of such a disaster. Andra made an active contribution to the work of this CODIRPA by developing a modular disposal concept, based on existing solutions in the CSA and CIRES and suitable for very large volumes of low and intermediate level radioactive materials ($>$100,000 m^3). It is also capable of being rapidly deployed to the site of the accident.

In this arrangement, the requirements concerning the local geological barrier are less strict, even if it must still offer mechanical stability, but the other performance criteria, notably soil permeability, required to protect the environment, are taken up by the engineered barriers built directly in-situ.

With rapid deployment of these solutions in the vicinity of the accident site, the contaminated materials can thus be kept within a delimited, protected and containing space, minimising the risk of natural disseminaion of isotopes into the environment and making any attempted retrieval highly improbable or even impossible.

7.8 Conclusions

Our societies are faced with constantly rising exposure to threats from multi-faceted, stateless terrorist organisations and it is their duty to analyse the potential risks and anticipate the deployment of countermeasures tailored to the new situations made possible by the technological, social and economic changes we have been experiencing for more than 30 years.

With the world's second largest nuclear power plant fleet, France took stock at a very early stage of the potential threat linked to the use of radioactive materials, even with low levels of ionisation, considered by some to be valueless waste with no possibility of recycling, but which could nonetheless become a frightening means of exerting pressure by organisations seeking to destabilise States through terror.

Ninety percent of the volume of the materials to be considered is essentially low and very low level. Its packaging by the waste producers (Areva, EDF, CEA) upstream of preparation for final disposal, significantly limits any interest in acquiring such materials. The disposal conditions on the Andra sites in reinforced concrete vaults, closed by a cover several metres thick, then provides security both in terms of protection of the environment and the population, and with regard to limiting the risk of intrusion.

The recycling of spent fuels gives France a significant technical edge, because the volume of high-level final waste is small, as well as an economic advantage, through the use of the uranium and the plutonium generated. Furthermore, the process of vitrification of the residual fission products in a glass matrix makes any attempt to misappropriate the encapsulated material highly improbable, because the means needed to extract it would demand resources on a very large scale. Finally, France intends to dispose of this waste at a great depth, 500 m below the surface, where the natural geological barrier would prevent all access.

Finally, France has developed a strategic plan for management of post-accident situations such as that which unfortunately struck Japan in 2011. For this, Andra created original final disposal concepts capable of being rapidly deployed and accepting very large volumes of materials. These disposal conditions would not only protect the populations and the environment, but would also make it very hard to extract material for malicious ends.

Although the fight against Evil is something that is destined to continue eternally, France and Andra in particular remain vigilant with regard to all and any possible attempts to misappropriate what is considered to be radioactive waste, whatever its activity level. In so doing, they are developing innovative solutions optimising both environmental protection and security.

Part II
Combating Nuclear Terrorism

Chapter 8
NATO Nuclear Policy, the Ukraine Crisis, and the Wales Summit

Jeffrey A. Larsen

Abstract NATO's 2010 Strategic Concept called upon the Alliance to carry out three core tasks: collective defense, crisis management, and cooperative security. Since the beginning of the Alliance each of those tasks has been emphasized at different times. Following the 9/11 attacks NATO joined the United States in becoming an expeditionary alliance, focusing on crisis management and collective security through robust partnership programs and out of area conventional military operations around the globe. Deterrence policy, and nuclear matters in particular, were de-emphasized. Indeed, some argued that there was no longer any requirement for NATO to have a nuclear strategy, nor US warheads stationed in Europe. With the end of its mission in Afghanistan at the end of 2014, the member states were already facing the challenge of determining whether the next version of NATO would continue to play the same expeditionary role. Some of the new member states in Central Europe, however, were hoping to return to an emphasis on Article 5 of the Washington Treaty: collective security and territorial defense. Russia's aggression in early 2014 against Ukraine made that a much more immediate concern for the Alliance as a whole, and gave greater import to the NATO summit later that year.

This chapter considers NATO nuclear policy in light of recent events and decisions from the September Wales Summit. It reviews the significant diminishment of the nuclear mission since the end of the Cold War, and addresses possible changes or updates to that mission given Russia's sudden change from strategic partner to potential adversary. It emphasizes the importance of deterrence, which includes nuclear weapons, missile defenses, and conventional forces, to the underlying mission of the Alliance, which is the security of the member states, their populations, and their strategic interests.

J.A. Larsen (✉)
NATO Defense College, Roma, Italy
e-mail: j.larsen@ndc.nato.int

S. Apikyan, D. Diamond (eds.), *Nuclear Threats and Security Challenges*, NATO Science for Peace and Security Series B: Physics and Biophysics,
DOI 10.1007/978-94-017-9894-5_8

8.1 NATO Nuclear Policy

The member states of the North Atlantic Treaty Organization (NATO) have been protected by nuclear weapons for three generations, since the first American atomic artillery projectiles were placed in Central Europe in 1953. Since then the Alliance has added two additional nuclear weapons states, has created elaborate multinational planning, basing, security, and employment processes, including roles for host nations teamed with US warheads, and saw its atomic arsenal grow to more than 8,000 US, British, and French warheads stationed in the theater in the 1970s. Since then, those numbers have diminished significantly, until today there are but a few hundred warheads left in Europe. Yet, according to the NATO web site, it remains true today that

> The supreme guarantee of the security of the Allies is provided by the strategic nuclear of forces the Alliance, particularly those of the United States; the independent strategic nuclear forces of the United Kingdom and France, which have a deterrent role of their own, contribute to the overall deterrence and security of the Allies.[1]

The Alliance points out that "dramatic changes in the Euro-Atlantic strategic landscape brought about by the end of the Cold War were reflected in the Alliance's 1991, 1999 and 2010 Strategic Concepts"[2] and in the 2012 Deterrence and Defense Posture Review. At the same time, the United States has produced a series of national security strategies and nuclear posture reviews (the latest of which was released in 2010) that also reflect the dynamic international security environment and the changing role for nuclear weapons in deterrence policy.[3]

NATO also likes to highlight its reduced reliance on nuclear forces which has been manifested in "steady and very significant reductions in the number of systems, overall weapon numbers and readiness levels since the end of the Cold War." This has had an impact on operational readiness levels, as well: "NATO no longer has standing peacetime nuclear contingency plans, and NATO's nuclear forces do not target any country."[4]

All recent documents stress the point that as long as there are nuclear weapons in the world, NATO will remain a nuclear Alliance. This does not mean that US weapons will remain stationed in Europe, but with three nuclear member states it cannot but be a nuclear alliance. And as NATO strategy highlights, "Deterrence, based on an appropriate mix of nuclear and conventional capabilities, remains a core element of NATO's strategy, even though the circumstances in which any use of nuclear weapons might have to be contemplated are extremely remote."[5]

---

[1] NATO home page at http://www.nato.int/cps/en/natohq/topics_50068.htm

[2] Ibid.

[3] *Nuclear Posture Review Report* (Washington: Department of Defense, April 2010), at http://www.defense.gov/npr/docs/2010%20nuclear%20posture%20review%20report.pdf

[4] NATO Home Page.

[5] Ibid.

There are significant political and bureaucratic implications to the members of a nuclear alliance.

> Political oversight and control are the cornerstones of NATO's nuclear posture and are shared among member countries. NATO members agreed to ensure the broadest possible participation of Allies in collective defence planning on nuclear roles, in peacetime basing of nuclear forces, and in command, control and consultation arrangements. Within the NATO HQ structure, the Nuclear Planning Group (NPG) provides a forum in which nuclear and non-nuclear Allies alike (except France, which has decided not to participate) engage in the development of the Alliance's nuclear policy, and in decisions on NATO's nuclear posture.[6]

New members are considered full members of the Alliance in all respects, including the requirement to commit to the Alliance's policy on nuclear weapons.

8.2 The Wales Summit

The 2014 Ukrainian crisis forced the North Atlantic Alliance to reconsider the appropriate balance between the three goals established in NATO's 2010 Strategic Concept: collective defense, crisis stability, and cooperative security.[7]

The emphasis over the past 20 years has been on the latter two of those three policy objectives—crisis stability and cooperative security. Collective defense has been increasingly neglected by the member states. This seemed like a reasonable approach at the time for several reasons. The major threat from the East during the Cold War seemed to have evaporated; the call for out of area operations around the globe increased dramatically after the end of the Cold War in the early 1990s (as seen most evidently in Bosnia, Afghanistan, and Libya); and the lack of a immediate "peace dividend" after the fall of the Berlin Wall made all member states anxious to reduce military spending wherever feasible. All of these efforts meant that in a smaller military that emphasized certain objectives, the other pillars necessarily had to suffer—and that meant collective defense under the Washington Treaty.

Collective defense was, of course, the original raison d'etre for the Alliance when it was created in 1949. NATO's purpose was to protect the territorial integrity and sovereignty of the member states in Western Europe, with military forces-in-being facing a massive Soviet conventional threat in Eastern Europe. When it became clear early on that the allies were unable and unwilling to provide the necessary conventional forces to meet that threat, the Alliance turned to nuclear deterrence and the threat of first use of atomic weapons in order to stem any Soviet or Warsaw Pact aggression. This ultimate guarantee of a state's integrity was enshrined in the US extended deterrence commitment to its allies.

---

[6]Ibid.

[7]Parts of this section appeared in NDC Research Paper, "The Wales Summit and NATO's Deterrence Capabilities: An Assessment," by Jeffrey A. Larsen, November 2014.

This guarantee worked for many decades. As with any deterrence relationship, it is impossible to prove whether deterrent forces or political commitments were most instrumental in preventing the outbreak of war on the European continent for more than 50 years. Yet since the end of the Cold War there have been increasingly frequent signs from some member states that they questioned the strength of that commitment. This concern over the credibility of Article 5 was most often heard from the new member states found in Central and Eastern Europe, those that were in the Warsaw Pact during the Cold War but happy to leave the embrace of communism and join the West. Even before the 2014 Ukraine crisis several of the East European member states were raising such concerns over the neglect of collective defense, which was the primary reason most of them joined the Alliance. These concerns, however, met with little resonance in the West.

A commonly heard refrain from Alliance insiders is that "we do not want to return to the Cold War." At first blush, this is a seemingly reasonable statement. Yet, cold war is certainly better than hot war, which could be the result if the Alliance lets down its guard in the face of a great power that has the capability, if not yet the confirmed will, to pursue an aggressive foreign policy. As in the case of conventional forces, this implies a need for well thought-out responses, including military preparedness. It is important to remember that the West only gets one vote in whether it returns to a cold war; it must also respond to how its adversary decides to proceed.

The Alliance did respond to the Ukraine crisis, although in a quite measured and limited way. It deployed some modest conventional forces to the Baltic States and Poland in an effort to restore confidence in its Article 5 commitments on the part of Allies and the potential adversary. The United States also deployed two B52 strategic bombers to the United Kingdom and to an exercise in the Baltics during the height of the tensions. It is unclear whether this was meant to be a signal to Russia. The combination of these activities, as well as statements from senior leaders of the Alliance, make it clear that the focus on the three core pillars of Alliance security—collective defense, cooperative security, and crisis management—has shifted to an enhanced emphasis on collective defense, at least for the short term. This may imply the reversal of a two-decade long trend that witnessed regular pronouncements about the importance of Article 5, while at the same time some states were reducing the forces necessary to conduct that mission.

The Wales communiqué repeats the standard NATO liturgy that the Alliance relies on an "appropriate mix" of forces: conventional, missile defense, and nuclear. But upon closer examination, there appears to be a problem with that equation and its levels of emphasis at the summit. The Alliance's conventional forces are not pre-positioned in the right place in the right numbers to meet the current threat and assure those member states that feel threatened. This reflects, in part, conscious decisions by the Alliance over the past two decades to restructure, reduce numbers, reduce readiness, and create a different kind of military force from the kind it had during the Cold War. It also reflects the commitments made to Moscow at the time of the first NATO enlargement that the Alliance would not station significant forces in the new member states.[8] This was meant to be a confidence building measure for Russia. Yet it is a political commitment, not a legally binding international agreement. If circumstances change, the Alliance has every right to revisit this commitment.

For 20 years, Alliance communiqués have minimized any threat in Europe, calling Russia a "strategic partner" with special benefits, including its own forum for direct communication with the Alliance (the NATO-Russia Council). Members derived a peace dividend through the overall diminution of military forces. At the same time, the Alliance created a different type of military, one better-suited to out-of-area operations that required greater responsiveness, greater flexibility, a smaller footprint, and ease of transportation. This meant fewer main battle tanks, armored vehicles, medium and long range missiles, medium and heavy bombers—in sum, none of the old-fashioned expensive equipment that the Alliance previously had to keep on hand in large numbers in Central Europe to deter or respond to Soviet aggression. That means there is very little of that kit left in Europe at a time when the Alliance might wish it had some to deploy to the weak spot on its perimeter.

The second leg of the appropriate mix of forces is missile defense. NATO continues to emphasize its requirement for a robust layered missile defense system, and rightly so. Given the continuing push for a nuclear weapons capability in Iran, and the unsettled nature of the entire region of the Middle East and North Africa, it is appropriate that a security alliance would want to protect its citizenry from the possible threat of missile attack. The European Phased Adaptive Approach is

---

[8] See, for example, "NATO's Eastward Expansion: Did the West Break Its Promise to Moscow?" Uwe Klußmann, Spiegel Online International, 26 Nov 2009, at http://www.spiegel.de/international/world/nato-s-eastward-expansion-did-the-west-break-its-promise-to-moscow-a-663315.html. For the counterargument, see Michael Rühle, "NATO Enlargement and Russia: Die-Hard Myths and Real Dilemmas," NDC Research Report, NATO Defense College, 15 May 2014, available at http://www.ndc.nato.int/research/series.php?icode=3; "NATO Enlargement and Russia: Myths and Realities," *NATO Review* online at http://www.nato.int/docu/review/2014/Russia-Ukraine-Nato-crisis/Nato-enlargement-Russia/EN/index.htm; and Mark Kramer, "The Myth of a No-NATO-Enlargement Pledge to Moscow," *The Washington Quarterly*, April 2009, at http://dialogueeurope.org/uploads/File/resources/TWQ%20article%20on%20Germany%20and%20NATO.pdf

According to the NATO-Russia Founding Act of 1997: "NATO reiterates that *in the current and foreseeable security environment*, [emphasis added] the Alliance will carry out its collective defence and other missions by ensuring the necessary interoperability, integration, and capability for reinforcement rather than by additional permanent stationing of substantial combat forces. Accordingly, it will have to rely on adequate infrastructure commensurate with the above tasks. In this context, reinforcement may take place, when necessary, in the event of defence against a threat of aggression and missions in support of peace consistent with the United Nations Charter and the OSCE governing principles, as well as for exercises consistent with the adapted CFE Treaty, the provisions of the Vienna Document 1994 and mutually agreed transparency measures. Russia will exercise similar restraint in its conventional force deployments in Europe." See *Founding Act on Mutual Relations, Cooperation and Security between NATO and the Russian Federation signed in Paris, France*, at http://www.nato.int/cps/en/natolive/official_texts_25468.htm

well underway and on track, with interim operational capability announced at the Chicago Summit in 2012. Plans continue to deploy ground-based AEGIS ASHORE batteries in the Balkans next year, and to continue to build improved command and control and early warning installations across the southern and eastern reaches of the Alliance. AEGIS ships are also now deployed permanently in the Mediterranean. All these capabilities, while impressive, would have little applicability against the type of threat that Ukraine had to face this spring: so-called hybrid warfare, an approach not dissimilar to low intensity warfare, involving guerrillas, non-uniformed militias, and conventional military equipment, with no need for missile defense.

That leaves the Alliance with the third leg of the deterrence triad: nuclear forces. Russia's national security doctrine since the end of the Cold War now resembles NATO's old Cold War strategy: first use of nuclear weapons if necessary to control the escalation ladder (called, paradoxically, "de-escalation" in Russian papers).[9] Russian nuclear doctrine also calls for early nuclear use on a conventional battlefield. Indeed, some analysts report that however good Russia may be at hybrid warfare and signaling via snap exercises, its conventional capabilities at the operational level remain weak. That would imply that in a conventional conflict which appeared not to be going their way, one might anticipate a jump to nuclear use. As in the scenario about returning to cold war, in a conflict the Alliance similarly gets but one vote in determining whether the fighting remains conventional.[10] Is NATO prepared to consider this possibility?

Given this combination of factors—the diminishment of conventional capabilities, the major cuts in the European missile defense program plan, the need for greater reliance today on collective defense, the overall diminishment of conventional military readiness by all member states in NATO since the global economic crisis began, and the call for greater levels of security reassurance by some of NATO's eastern member states—one would expect to see an *increased* level of emphasis on the Alliance's remaining pillar: nuclear forces. This, one could argue, would be especially true in the short term, as the Alliance found itself having to respond to aggressive behavior by a major nuclear state on its borders. This increased reliance might be seen, for example, in any of the following steps:

- a halt to any further planned reductions in nuclear forces
- signaling to the adversary, whether by way of exercises, nuclear force movements, or simple announcements as to the increased reliance on nuclear deterrence
- an increase in replacement decisions for the dual-capable aircraft fleets of many nations that are nearing retirement
- a reversal of the Prague Vision and its call for nuclear disarmament, which the Alliance took up in its DDPR

---

[9]For a summary of Russian nuclear policy today, see Nikolai Sokov,"Why Russia Calls a Limited Nuclear Strike 'De-Escalation,'" *Bulletin of the Atomic Scientists*, 13 March 2014, at http://thebulletin.org/why-russia-calls-limited-nuclear-strike-de-escalation

[10]See Larsen and Kartchner [1].

- an increase in the frequency or visibility of Nuclear Planning Group meetings
- more formal and pointed US commitments to the extended deterrence guarantees it provides its NATO allies, such as it recently did for Japan.

In general, this would imply a return to some of the old ways of dealing with Russia during the Cold War, when nuclear deterrence was understood and brandished as a legitimate source of Alliance power.

Yet none of those things has happened—at least not in a structured, public manner designed to reassure the new member states and deter Moscow. There has been very little public debate over the nuclear implications of the Ukraine crisis. Nor, according to sources within NATO and SHAPE Headquarters, has there been any apparent increase in discussions or meetings within the Military Committee, Nuclear Planning Group, or North Atlantic Council regarding nuclear policy or its myriad aspects (including military requirements, nuclear forces and their delivery means, command and control procedures, and so on); no reversal of national objections to Dual-Capable Aircraft (DCA) purchases; no increased appetite among any of the member states for the consideration of nuclear deterrence as a reasonable or legitimate military approach to the problem. This despite calls over the past 20 years for an appropriate mix of forces, including nuclear forces.

Surprisingly, there has been very little reaction within NATO to the requirement for re-balancing the three pillars, and no desire to emphasize nuclear capabilities for reasons both financial and political. It appears the Alliance is once again more focused on "not rocking the boat" than on an informed public discussion over nuclear matters.[11]

8.3 The Role of Nonstrategic Nuclear Weapons in NATO Policy and Alliance Strategy[12]

If thinking about a return to cold war is unwelcome within the Alliance, thinking about any role for nuclear weapons is anathema. The perceived value of nuclear weapons has dropped significantly within Alliance circles over the past 20 years. The few short paragraphs in the wales communiqué dealing with this subject are simply cut and paste repeats of nuclear language that has been used in such reports for over a decade. Furthermore, the remaining wording for the first time dropped all reference to US non-strategic nuclear warheads stationed in Europe. However

[11]Not rocking the boat, or letting sleeping dogs lie, are two commonly heard aphorisms in debates regarding NATO nuclear policy. These are not new perspectives. For an explanation of the use of such terminology, see Jeffrey A. Larsen, *The Future of US Non-Strategic Nuclear Forces and Implications for NATO: Drifting Toward the Foreseeable Future*, final report of the 2005 Manfred Wörner Fellowship Program, 31 October 2006, at http://www.nato.int/acad/fellow/05-06/larsen.pdf

[12]Potions of this section were previously published by Larsen [2].

the decision was made on what to include in the communiqué, from the results it is obvious that the Alliance chose to minimize this element of Alliance policy, at least for public consumption. NATO needs to have a dialogue, but that requirement has been discouraged by some members who fear the possible spillover effects of a debate over the future of nuclear forces, or are genuinely anti-nuclear but do not wish to be seen as opposing the big three nuclear powers, or worry that discussion over nuclear policy would appear provocative to Moscow. Granted, there are certainly valid arguments that can be made for taking this non-confrontational approach and minimizing any discussion on nuclear matters. But in the midst of a crisis involving the world's second largest nuclear power, it is striking to note how little was said about nuclear deterrence capabilities—at least on the western side of the crisis.

There is, however, an alternative, more optimistic hypothesis. It is also possible that the value of NATO's Article 5 commitments and its existing deterrent capabilities remain strong, as was shown by Russian inaction, the fact that it has not yet attempted to make trouble with the Baltic States or Poland. This line of argument would suggest that Moscow does still respect the Alliance's collective defense capabilities and political will. Nuclear weapons remain the ultimate insurance policy for the West. The size of that policy may be less important than its existence. If that is the case, the restrained language in the Wales summit communiqué was quite sufficient as a subtle reminder of those red lines.

Some analysts have questioned whether the United States needs to continue to deploy nuclear weapons in Europe. After all, they say, it has been more than 20 years since the collapse of the Warsaw Pact and demise of the Soviet Union; surely these weapons are no longer needed to ensure peace in a Europe whole and free. Yet official NATO policy still views nonstrategic nuclear weapons as a deterrent to any potential adversary, and they also serve as a link among the NATO nations, with shared responsibility for nuclear policy planning and decision-making. They also still serve as a visible reminder of the U.S. extended deterrent and assurance of its commitment to the defense of its allies. But as the Congressional Research Service (CRS) has written, if the United States and its allies agree that this assurance can be provided with either conventional capabilities or strategic nuclear weapons, the need for forward basing in Europe may diminish. Some argue that because these weapons play no military or political role in Europe, they no longer serve as a symbol of alliance solidarity and cooperation. Others, however, including some officials in newer NATO nations, have argued that U.S. nonstrategic nuclear weapons in Europe not only remain relevant militarily, in some circumstances, but that they are an essential indicator of the U.S. commitment to NATO security and solidarity.[13]

[13]Woolf, *NSNW*, p. 24. Also see Simon Lunn, Chapter 1, "NATO Nuclear Policy—Reflections on Lisbon and Looking Ahead," forthcoming in *The NTI Study on Nuclear Weapons and NATO*, draft 6 May 2011; and Paul Shulte, "Is NATO's Nuclear Deterrence Policy a Relic of the Cold War?" Policy Outlook, Carnegie Endowment for International Peace, 17 November 2010.

Political trends in Europe may accelerate these changes to U.S. forward deployments of non-strategic nuclear weapons. Prior to the 2014 Ukraine crisis, a number of factors appeared to be driving the Alliance toward ending the nuclear mission—or at least removing the remaining U.S. warheads.[14] Russia has had a strategy for years that uses diplomatic and political pressure against the United States and its allies in Europe to remove the remaining U.S. warheads. The European allies are suffering from two generations of military and particularly nuclear malaise, and seem unwilling to continue this effort given the political and economic costs of buying a next generation of DCA capable aircraft. The technical expiration of the service life of both the warheads and their delivery systems means that a decision has to be made to continue the NATO nuclear deterrence mission. Organizational changes in recent years within the U.S. Office of the Secretary of Defense, and NATO's Nuclear Policy Directorate and SHAPE Nuclear Planning office, have marginalized the nuclear mission to a considerable extent. The U.S. Air Force has never liked this mission, and has little in the way of U.S.-based DCA capabilities to back up mission requirements in Europe or Asia. For example, recognizing the changed international security environment, as well as these organizational and operational changes, several years ago NATO increased its response time for alert aircraft from minutes, as it was during the Cold War, to weeks.[15]

There are various schools of thought within elite NATO circles as to the value, role, and future of NSNW in Europe. These perspectives range from traditional supports, to believers in selective engagement, to proponents of arms control, to disarmament advocates.[16] At a meeting in Tallinn, Estonia in April 2010, NATO's foreign ministers sought to balance the views of those nations who sought the removal of the weapons with those who argued that these weapons were still relevant to their security and to NATO's solidarity. At the conclusion of the meeting, Secretary of State Hillary Clinton said that the United States was not opposed to reductions in the number of U.S. nuclear weapons in Europe, but that the removal of these weapons should be linked to a reduction in the number of Russian nonstrategic nuclear weapons. The foreign ministers also agreed that no nuclear weapons would be removed from Europe unless all 28 member states of NATO agreed.[17]

Others have raised the question whether the United States and NATO might benefit from the removal of these weapons from bases in Europe for reasons of

[14]For representative arguments about the European social situation and NATO nuclear policy, see Jeffrey A. Larsen, "Future Options for NATO Nuclear Policy," Issue Brief, The Atlantic Council, August 2011, available at http://www.acus.org/publication/future-options-nato-nuclear-policy; Lunn, "NATO Nuclear Policy;" and Bruno Tertrais, "Extended Deterrence: Alive and Changing," *The Interpreter*, Lowy Institute for International Policy online, 2 February 2011, available at www.lowyinterpreter.org/post/2011/02/02/Extended-Deterrence-Alive-and-Changing

[15]"NATO's Nuclear Forces in the New Security Environment," *NATO Issue Brief*, NATO Headquarters, Brussels, June 2004.

[16]Professor David Yost has recently reviewed these distinctive views and outlined the key questions under debate. Yost [3], pp. 1401–1438.

[17]Woolf, *NSNW*, p. 25.

safety and security, as well as cost-saving. Some analysts have suggested that, in response to these concerns, the United States might consolidate its nuclear weapons at a smaller number of bases in Europe. According to another study, officials at U.S. European Command have argued that weapons deployed outside of Europe could be just as credible as forward deployed weapons as a deterrent to attack on NATO.[18] In fact, in recent years some observers argued that reducing or eliminating U.S. nuclear weapons in Europe would not only address the Air Force's operational and security costs associated with their deployment, but also could serve as a signal to Russia of NATO's intentions to address Russia's perception of the threat from NATO.[19] Of course, much of that argument has been suspended as a result of Russia's aggression against Ukraine in 2014.

8.4 The Way Ahead

What are the implications for this apparent lack of interest in even discussing the nuclear leg of NATO deterrence strategy? First, this lack of concern cannot but undermine reassurance to those member states in Eastern Europe that joined the Alliance in large part to secure the Article 5 security guarantee for their country against just this scenario—of a revanchist and revisionist Russia once again becoming a threat. We should not expect much in the way of new initiatives or new capabilities in the Alliance's nuclear forces in the future, nor any reversal of state decisions to reduce their F-35 orders. And while some attention is being given to a renewed interest in collective defense by certain of NATO's political leaders, the topic was not given significant discussion at the Wales Summit.

Second, The Alliance, according to some analysts, is facing confusion and disunity over its strategic direction. Such "strategic incoherency" was brought about by rampant enlargement; by more than a decade of expeditionary operations, which may have helped alliance interoperability but did little to create a sense of purpose or direction for the members; and by a crisis in Ukraine that, while putting a spotlight on the strategic shortfall, did not create enough sense of urgency to cause states to call for corrective action.

Whatever else Russia may be to Europe, it remains a huge market for European goods, and a major source of energy for the countries of Europe, the EU, and NATO. This combination of long-term energy dependency and capitalism will likely override any short term security issues that don't directly affect a NATO member state. As others have pointed out, the business of the West is business, and those businesses want things to return to normal as quickly as possible.

For the time being, any change to the status quo is unlikely. The current position on NATO's nuclear policy was difficult to achieve at the Lisbon summit and in the

[18]Woolf, *NSNW*, p. 26.

[19]Ibid.

follow-on DDPR. No member state wants to reopen the debate unnecessarily. This was shown once again by the paucity of language on deterrence in the 2014 Wales Summit communique.

Indeed, it is quite likely that after a short pause, assuming Moscow does not do anything else to antagonize the West and inflame world opinion, the Alliance members will revert to their more immediate concerns, focusing primarily on how to achieve even greater efficiencies in their defense budgets after withdrawing from the 14 year operation in Afghanistan. The trend of the past 25 years in NATO's nuclear forces may then resume its vector toward zero US weapons based in Europe. All the arguments heard for the past two decades on why the Alliance should eliminate its European-based nuclear forces will return to the fore. Life will go on as it did prior to the mess in Ukraine, with the Alliance once again focusing on small-scale operations it can use for training opportunities, capacity building, host nation support, smart defense, and the like, while providing only benign neglect to its collective defense responsibilities.

The impact of the Ukraine crisis on NATO-Europe has therefore been little more than a pebble cast into the pond of international security affairs. It created some ripples, causing some of the toy boats to bob about, but in the absence of any additional rock tossing the waves will soon subside, and things will return to normal. Or so Europe wants to believe.

References

1. Larsen J, Kartchner K (eds) (2014) On limited nuclear war in the 21st century. Stanford University Press, Palo Alto
2. Larsen JA (2012) Chapter 15: The role of non-strategic nuclear weapons: an American perspective. In: Stuart D, McCausland J (eds) Tactical nuclear weapons and NATO. Strategic Studies Institute, US Army War College, Carlisle
3. Yost DS (2011) The U.S. debate on NATO nuclear deterrence. Int Aff 87(6):1401–1438

Chapter 9
Risk Informing Security at the US Nuclear Regulatory Commission

Joseph D. Rivers

Abstract The US Nuclear Regulatory Commission (NRC) has efforts underway to better risk inform security regulations. The NRC has conducted two workshops to obtain stakeholder feedback on how to accomplish this. Topics at the last workshop included an in depth look at risk methodologies/approaches, the likelihood of initiating events, modeling and simulation tools, cyber security, and the using the attractiveness of special nuclear material to an adversary to inform NRC's graded security program. In addition, the NRC staff is responding to recommendations from a NRC Risk Management Task Force that concluded in 2012. Among its recommendations was to better communicate between safety and security to allow both disciplines to leverage off each other. This paper will summarize the current initiatives to better risk inform security and provide a detailed overview of NRC's efforts to consider the attractiveness of special nuclear material to a potential adversary in the development of a graded security program. In particular, this effort has identified levels of dilution as a key factor that could be used to adjust security requirements.

9.1 Background

Over the last half dozen years, the NRC has worked to identify ways to better risk inform its security regulatory process. The NRC's safety regulatory process has a strong tradition of being risk informed. In order to identify what might be possible, the NRC engaged in a number of activities. The NRC conducted a workshop hosted by Sandia National Laboratories in 2010, which brought together safety and security risk experts form the government, national laboratories, and universities. Commissioner George Apostolakis led a Risk Management Task Force from 2011 to 2012. The NRC conducted a second workshop in 2014 hosted by the Institute of Nuclear Materials Management (INMM).

J.D. Rivers (✉)
Nuclear Regulatory Commission, Washington, DC, USA
e-mail: joseph.rivers@nrc.gov

S. Apikyan, D. Diamond (eds.), *Nuclear Threats and Security Challenges*,
NATO Science for Peace and Security Series B: Physics and Biophysics,
DOI 10.1007/978-94-017-9894-5_9

The Sandia Workshop brought together security and safety risk experts from government, national laboratories and universities. The first half of the workshop consisted of presentations on the topic of risk informing security. Workshop participants were split into groups during the second half of the workshop to identify areas/topics that might be fruitful for focusing efforts in the future. The workshop identified six possible areas that might be pursued to better risk inform security:

- Uncertainty of initiating events
- Use of simulation tools
- Collaboration between safety and security professionals
- Cyber security
- Need for improved metrics
- Development of a demonstration project like WASH 1400.

The Risk Management Task Force was tasked with developing a strategic vision and options for adopting a more comprehensive, holistic, risk-informed, performance-based regulatory approach. The task force identified recommendations for the major program areas within the NRC's regulatory program. However, the recommendations tended to be predominantly safety oriented. The limited discussion of security focused on the need to make sure that terminology used in the safety and security disciplines was better understood. In particular, the need was identified to produce a glossary to identify how terminology used in the two disciplines was either similar or different.

In February 2014, the INMM hosted the Risk Informed Security Workshop in Stone Mountain, Georgia. The workshop was organized as a follow-on to the Sandia Workshop, providing an opportunity to dig deeper into areas that had been identified at the earlier workshop. In addition to some background presentations, there were six panel discussions:

- Safety/Security Risk Approaches
- Material Attractiveness
- Likelihood of an Initiating Event
- Vulnerability Assessment Simulation Tools
- Cyber Security
- Risk Management Approaches.

In general, the workshop participants believed that most of the areas warranted further investigation. The most challenging topic would be the likelihood of an initiating event. In the near-term, the most benefit was perceived to be possible from having interactions between safety and security risk professionals.

9.2 Current Activities

There are a number of current activities that are addressing to some extent the possible improvement in risk informing security. These activities include industry initiatives, internal NRC working groups, rulemaking activities, and international projects.

The Nuclear power plant industry is attempting to develop a Risk Prioritization Initiative (RPI). This effort is focused on developing a process for plants to prioritize projects at a facility based on measures of risk. The intent is to assess the project's ability to reduce risk, considering safety, security, emergency preparedness, radiation protection, and reliability. Projects that rate highest would be prioritized and initiated earlier than others. The RPI is in early stages, with the Nuclear Energy Institute having developed draft guidance and conducted pilot exercises at six nuclear power plants. The safety and reliability portions of the process are much more mature than the other three disciplines. The NRC staff will be providing a paper to the Commission next March related to this initiative, with recommendations for further work in this area.

The Risk Management Regulatory Framework Working Group was established to respond to the recommendations made by the Risk Management Task Force. This interoffice working group initially began work to address the entire scope of the task force recommendations, as well as to develop a policy statement on defense-in-depth. Earlier this year, the scope of the working group's efforts was reduced to just power reactor safety. As a result, the security focus on this effort is to monitor and make sure that what is developed for reactor safety will not adversely impact future related work in security.

The nuclear power plant industry has recently started investigating the possible use of modelling and simulation tools for the conduct of vulnerability assessments. These tools have the possibility of supporting training, management decisions, development of protective postures, and regulatory actions. The focus for NRC staff is to determine how the use of these tools might be incorporated into the regulatory process. In order to do this, the staff will need to be trained to better understand how these tools work, as well as to engage with a wide variety of stakeholders, to include vendors, other government agencies and users of the tools. In addition, the NRC staff will observe how the plants and their security systems are modelled and populated with data. Although the NRC is very early in the process, the use of these tools should offer a more structured approach in assessing possible changes in security programs at facilities.

The NRC has been working for several years to develop a more graded security program for special nuclear material (SNM). Rather than assuming that all uranium and plutonium is equally attractive to a potential adversary, the NRC has studied how the chemical and physical forms may impact the attractiveness of the SNM. If the SNM is more attractive, the adversary is more likely to attempt to steal the SNM and construct an improvised nuclear device (IND). Many such approaches have been investigated over the years, but they generally produce attractiveness approaches that would be too complex to incorporate into a regulatory regime.

In the end, an approach was developed that focused on attractiveness being represented by levels of dilution. In principle, the more dilute the SNM, the less attractive the SNM would be to an adversary. Dilute material requires larger volumes and masses to be acquired, then the dilute material must undergo processing to be placed into a weapons usable form. In the end, this provides a level of defense-in-depth, as authorities would have more time to locate stolen material if it requires processing prior to use.

In order to provide risk insights into this possible approach to grading security requirements, the NRC tasked Los Alamos National Laboratory to develop a logic model to assess the likelihood that an adversary could actually achieve yield. The model was developed using four modules: acquisition, processing, weapons design and engineering, and degradation. The acquisition model considers generic facilities with 32 potential scenarios for 12 classes of adversaries. Given the security measures that are employed, the likelihood of acquisition is assessed for varying quantities and forms for each of the adversary classes. The processing module identifies the necessary chemical and metallurgical process that are necessary to place the SNM into a weapons usable form. The module assesses the likelihood of success for each of the adversary classes, with assumptions made as to financial and technical capabilities of the adversary classes. Next, the weapons module assesses the likelihood of the adversary to be able to design and construct an actual IND, considering a variety of designs. This module assesses the likelihood of adversary success across a range of yields. Finally, the yields are adjusted in a degradation module to reflect that adversaries have a wide range of skill levels. The model allows protection elements to be varied to allow risk levels to be maintained relatively the same across all quantities and levels of dilution for SNM.

Most efforts to risk-inform cyber security have focused on the associated consequences of a potential cyber attack. At this time, the NRC has regulations addressing cyber security at nuclear power plants. Nuclear power plant cyber controls must be in place to protect critical digital assets that could impact safety, security, or emergency preparedness systems. Currently, the critical digital assets with the highest possible consequences should be incorporated into the plants' cyber programs. The final step will be to address the remaining, lower consequence related critical digital assets. There are efforts at this time to identify lower levels of protection that would be warranted for these assets. In addition, the NRC is evaluating potential cyber consequences at fuel cycle facilities, research and test reactors, spent fuel storage facilities, and radioactive materials facilities to determine what if any cyber requirements should be imposed on them.

On the international level, the NRC staff is leading an International Atomic Energy Agency Coordinated Research Project on Nuclear Security Assessment Methodologies (NUSAM). This project is designed to develop a risk-informed, performance-based approach for assessing the effectiveness of security for a wide variety of plants and activities associated with nuclear and radioactive material. The project began in April of this year and is anticipated to conclude in April 2017, and includes participation from more than a dozen countries. The project will develop a detailed guide, providing the technical basis for the methodology. This guide would provide sufficient detail to conduct assessments at nuclear power plants and Category I fuel cycle facilities. The abridged guide would support the conduct of assessments at lesser facilities. In addition, the project is developing a number of case studies to demonstrate the methodology. These case studies include a nuclear power plant, an irradiator facility, transport of Category 1 quantities of radioactive material, a low-enriched uranium fuel fabrication facility, and a spent fuel storage facility. These case study documents will be useful in training and in demonstrating how to conduct an assessment at one of these facilities.

9.3 Next Steps

The NRC has found that workshops are very useful for identifying potential elements of a risk-informed security program. During 2015, the NRC will conduct or participate in three workshops to better risk-inform security: the INMM Reducing Risk Workshop, an American Nuclear Society (ANS)/INMM Workshop on Safety/Security Risk, and an INMM Workshop on Vulnerability Assessment Tools.

The INMM Reducing Risk Workshop will be the eighth in a series of workshops. It will be hosted by the George Washington University Elliot School of International Affairs in March. One of the four topics will be cyber security risk. The cyber security panel will focus on identifying approaches to more fully incorporate the concepts of risk into cyber security programs.

The ANS/INMM Workshop on Safety/Security Risk will be conducted in Sun Valley, Idaho in April. The previous two workshops had identified the need for risk experts from the safety and security communities to interact. This workshop will enable that to happen. Experts will be given the opportunity to discuss how their discipline uses risk insights, and discussions will be held to identify where safety approaches might be used in security and security approaches might be used in safety.

The INMM Workshop on Vulnerability Assessment Tools will be hosted by the INMM in Boston in September. After a brief discussion of the history of vulnerability assessments and the results of earlier workshops a number of topics will be provided, including:

- Discussion of verification, validation, and accreditation of software,
- Discussion of modelling challenges in nuclear security,
- Discussion of data resources,
- Discussion by vendors of their tools, using a common nuclear power plant as an example,
- Demonstrations by vendors of their products,
- Discussion by users of their experiences, and
- Panel discussion including vendors, users, and regulators.

The workshop should provide more clarity to how tools benefit nuclear security programs and what is necessary to incorporate vulnerability assessment software tools into a regulatory process.

9.4 Summary

The NRC is actively seeking opportunities to enhance its regulatory process using increased risk information. These opportunities are being developed by industry, internal to the NRC, and through the conduct of workshops. By enhancing the regulatory process with broader use of risk information, the regulatory process will become better structured and more appropriately graded.

Chapter 10
SCK•CEN'S Activities in the Field of CBRN Risk Mitigation: Practical Contributions to Improving Nuclear Security Governance

Klaas van der Meer, Carlos Rojas Palma, and Johan Camps

Abstract The Belgian Nuclear Research Centre SCK•CEN started activities in the framework of CBRN risk mitigation around 10 years ago. These activities were based on expertise on radiological and nuclear emergency management that was acquired over the years. The first activities focused on RN risks, in particular radiological dispersal devices or dirty bombs, followed by a more comprehensive study of the radiological risks and the countermeasures to be taken in the framework of a European FP6 R&D project called TMT Handbook. Present activities in CBRN risk mitigation of SCK•CEN have on the one hand a clear R&D component, mainly embedded in European research projects like e.g. the FP7 project CATO, and on the other hand a practical implementation component by leading or participating in several European projects (Centres of Excellence, Framework contracts for the Instrument for Stability) and providing advice and expertise to the national Belgian authorities. In this paper we will give an overview of all these activities and discuss some projects in more detail.

10.1 Introduction

Nuclear security has received an increasing amount of attention in the past decade. This attention has given rise to political action by various actors, of which the US and the EU are among the most important ones.

The term *nuclear security* in its strict interpretation is security of nuclear weapons and nuclear material that can be used for nuclear weapons. This was also the main scope of the Nuclear Security Summits, but the security of radioactive sources was also mentioned and gained some more attention in the course of time.

The instruments to improve nuclear security often have a broader scope than just nuclear. For instance, the European initiative of CBRN Centres of Excellence

K. van der Meer (✉) • C. Rojas Palma • J. Camps
Nuclear Research Centre SCK•CEN, Mol, Belgium
e-mail: kvdmeer@sckcen.be

S. Apikyan, D. Diamond (eds.), *Nuclear Threats and Security Challenges*, NATO Science for Peace and Security Series B: Physics and Biophysics,
DOI 10.1007/978-94-017-9894-5_10

considers not only the nuclear threat, but all possibilities for obtaining weapons of mass destruction (chemical, biological, nuclear).

The activities of SCK•CEN in the framework of improving nuclear security are mainly sponsored by European instruments and therefore do not always only focus on nuclear security. Some activities apply to chemical and biological threat reduction.

10.2 EC Instruments

The EC has created many instruments to support its policies. The instrument that is used to support nuclear security is the *Instrument contributing to Stability and Peace* (*IcSP*), formally known as the *Instrument for Stability* (*IfS*). The general scope of the IcSP is in the areas of crisis response, conflict prevention, peace-building and crisis preparedness, and addresses global and trans-regional threats, both man-made and natural. The IcSP has a short-term and a long-term component. One of the long-term components is the CBRN Centres of Excellence that aims at mitigating CBRN risks by improving CBRN security in regions all over the world.

Whereas the IcSP implements projects based on existing European expertise in CBRN, the EC also supports research to improve CBRN security. This research is sponsored via the framework projects in the past and now via Horizon 2020. Several Directorate-Generals are participating in this support, for instance DG Security and DG Home.

10.3 The Nuclear Security Summits (NSS)

The basis of the Nuclear Security Summits was laid in the 2009 Prague speech of President Obama, where he urged for action on the threat of misuse of nuclear material for terrorist's purposes. This resulted in the biannual organisation of the Nuclear Security Summits (2010 Washington, 2012 Seoul, 2014 The Hague). During these summits, actions are agreed upon to diminish the threat of the presence of nuclear and radiological material and these actions are discussed and evaluated during the next NSS. While nuclear material was the main focus at the start of the NSS, radioactive sources got more attention in the course of time. The next Nuclear Security Summit will be organised in the US. Examples of actions performed in this framework are the reduction of the use of Highly Enriched Uranium (HEU), reduction of HEU stockpiles, installation of detection equipment at borders to prevent illicit trafficking in nuclear and radioactive material and applying IAEA security standards in nuclear facilities.

10.4 Present Contributions of SCK•CEN to CBRN Risk Mitigation

SCK•CEN contributes in various ways to the reduction of CBRN risks. Its main contributions go via the EC Instrument contributing to Stability and Peace IcSP. Within this instrument, the CBRN Centres of Excellence play an important role for the SCK•CEN activities. Three projects for CBRN CoE are presently carried out, in which SCK•CEN plays a role. Furthermore SCK•CEN is leading a consortium for the so-called Expert Support Facility in lot 1 (CBRN Risk Mitigation). The EC can ask to provide on a short term experts for specific tasks in countries mainly outside the EU.

Another mayor contribution to CBRN risk reduction is done via research. SCK•CEN has participated or is participating in two R&D projects: TMT Handbook (FP5) and CATO (FP7).

In the framework of agreements within the Nuclear Security Summits SCK•CEN has removed nuclear material and transferred it to the USA.

10.4.1 CBRN Centres of Excellence Project 9: Lebanon

The main objective of the project was to enhance the national response plan in Lebanon. This enhancement dealt with legislative, organizational and practical implementation.

The Project Consisted of Five Work Packages

The first work package focused on the legislative and organizational part and included an assessment of the draft National Response Plan of Lebanon, followed by an assessment of the national response organization. The assessment was done following international examples and recommendations [1–4].

The second work package focused on organizational aspects, in particular on the development of Standard Operations Procedures (SOPs). Whereas the original idea was to assess available SOPs and provide recommendations for improvement, it appeared that SOPs did not yet exist. A procedure to develop SOPs, based on international guidance [5–11], was therefore written and a list of prioritised SOPs was identified. SOPs on this prioritised list should be developed first.

The third work package focused on the practical implementation of the response plan and SOPs. Guidance with respect to the development of table-top and field exercises has been developed for this purpose and provided to the beneficiary in the form of reports. Additionally a 5-day course has been delivered to representatives of the major stakeholders in national emergency planning and response in Lebanon. During this course vocational training was given related to more theoretical aspects of emergency response, while practical exercises were given to perform various table-top exercises in the CBRN field. Specific training in the tools HOTSPOT and RODOS was given to provide the knowledge and capacity to the Lebanese stakeholders to develop themselves new scenarios for table-top exercises. The Lebanese participants indicated that they preferred to organise a field exercise only after having participated as observer in a field exercise e.g. in Belgium. Since this was not possible within the present project, the performance of a field exercise was replaced by providing the European decision support system RODOS for radiological and nuclear emergencies. The Belgian national crisis centre was visited in the framework of the project.

The fourth work package dealt with the material aspects of emergency response. Lists of necessary CBRN first response equipment were provided. A specific demand from the beneficiary was a list for equipment for a field hospital, which was provided in the delivered report.

In the fifth work package worst-case accident scenarios were developed for the Dimonah reactor in Israel regarding radiological consequences for Lebanon. Calculations were performed with RODOS, a decision support system developed in the framework of European research programmes, and reported.

10.4.2 CBRN Centres of Excellence Project 16: North-Africa

The general objective of this project is to enhance the overall capacities of the countries to maintain effective control over radiological and nuclear materials and installations.

This Will Be Achieved by

Undertaking a review of the laws and regulations relating to radiological and nuclear security in each beneficiary country, and formulating recommendations for enhancing the existing legislation.

Reviewing the state of the physical protection measures at installations containing radiological and nuclear materials and providing recommendations, as necessary, for the upgrade of physical protection.

Assessing and formulating recommendations concerning the establishment of a central storage facility for radioactive material and orphan sources, including seized material.

Guiding the completion of the radiological and nuclear materials inventory, including the assessment of the inventory control procedures currently in effect.

Sharing best practices for the development of a national strategy for combating illicit trafficking in radiological and nuclear materials, and a national response plan to potential radiological or nuclear incidents.

All activities are made available to participating countries and shall be tailored in coordination with the individual countries in question.

It is expected that, by the end of the project, the countries will engage in the development of the national strategies and national response plans in line with the international recommendations and best practices. The very process will lead the countries to enhance their capacities in the radiological and nuclear security area and improve national nuclear security.

10.4.3 *CBRN Centres of Excellence Project 33: Central and East Africa and African Atlantic Façade*

Project 33 aims at technically implement in the countries Gabon, Mauritania, Morocco and Senegal, Democratic Republic of Congo, Kenya, Uganda and Burundi two components (Component 1 and Component 2).

The aim of **Component 1** is strengthening the CBRN national legal framework to comply with international obligations for mitigating CBRN risks.

The purpose of **Component 2** is the provision of specialized and technical training to enhance CBRN preparedness and response capabilities.

To this end, all countries are visited in the framework of component 1 by experts to evaluate the current legal framework with respect to CBRN security and to assess needs to improve CBRN security.

The needs assessment results in the provision of dedicated training curricula that will be given in the framework of component 2 to improve CBRN preparedness and response capabilities.

10.4.4 *Expert Support Facility (ESF) Lot 1 (CBRN Risk Mitigation)*

The ESF aims to provide quickly expertise, mainly in countries outside the EC. The following subjects are included in Lot 1 of this ESF:

- *Prevention of CBRN illicit trafficking and deceptive financial practices*
  - illicit financing of proliferation activities;
  - strengthening the capacity of relevant authorities involved in combating illicit trafficking and illicit financing;
  - illicit trafficking of CBRN materials, including in high seas;
  - export control of dual-use items.
- Support for bio-safety and bio-security
  - assessment of safety and security aspects related to the design, operations and management of bio labs;
  - coordination and exchange of technical expertise with key international organizations (WHO, Interpol etc.).

- Redirection of former weapon scientists and engineers
- Multilateral Nuclear Assurances

The project is active for the period 2014–2016. No requests for services have been received yet from the EC.

10.4.5 The R&D Project TMT Handbook in the Sixth Framework Programme (FP6)

The TMT Handbook project (Triage, Monitoring and Treatment) aimed at developing a practical guideline for first responders (fire brigade, civil protection, medical first response) to respond to radiological malevolent acts.

In this project scenarios of malevolent acts were developed (e.g. radiological dispersion device, radiation exposure device, food contamination) and possible consequences were estimated. On basis of these estimates countermeasures were developed and good practices established. These good practices dealt with e.g. establishment of safety and security perimeters, detection equipment for monitoring (persons, environment, goods) that should be available, triage procedures to allocate the available efforts most effectively, and patient treatment protocols for medical first responders. A training course was held for national authorities from EC Member States that provided the theoretical background for first response to radiological malevolent acts, and focused on the practical aspects of emergency response by providing a lot of training exercises, among which a large table-top exercise dealing with a significant radiological attack.

10.4.6 The R&D Project CATO in the Seventh Framework Programme (FP7)

The acronym CATO stands for CBRN, Architecture, Technologies & Operational Procedures and is designed to do just that – started in January 2012, CATO is an FP7 integration project, which develops and brings together a coherent toolbox of systems to allow better life cycle management of CBRN incidents. Twenty-five multi-disciplinary organisations from across Europe are involved in this research and development project, which is entering its final stages and will conclude in December 2014.

CATO has been tasked with developing a comprehensive Open Toolbox for dealing with CBRN crises due to terrorist attacks with CBRN substances or on facilities storing CBRN material. It is also building a dedicated community of user organisations and stakeholders concerned by CBRN Preparedness and Response, such as policy makers, ethics experts, emergency management agencies, healthcare responders, first responders, CBRN experts, equipment providers and system suppliers.

In the framework of CATO SCK•CEN has developed a software tool that aims at a fast analysis of the size and type of incident based on social networks communication, and it has developed a radiological dose assessment system for members of the public based on retrospective dosimetry with the use of Optically Stimulated Luminescence materials in cell phones or ID cards. SCK•CEN has also participated in several practical exercises to test the developed methods [12].

10.4.7 NSS

SCK•CEN has sent parts of its stockpiles of plutonium and HEU to the US to reduce the threat posed by the presence of this material. It furthermore implements the adapted Belgian legislation with respect to security.

10.5 Discussion

SCK•CEN has been involved in the CBRN Centres of Excellence from the start. It was awarded 2 (project 9 and project 16) of the first 19 projects that were managed by UNICRI on behalf of the European Commission. SCK•CEN is Consortium member of the Consortium led by France Expertise International (FEI) for project 33, a 3-years project of which 1 year has passed and it is Consortium Member of the Consortium led by Gesellschaft für ReaktorSicherheit (GRS) that won project 42. This is also a 3-year project that should start soon.

From the experience obtained from the first two projects it became soon clear during implementation that local ownership is essential for the success of such a project. This is even more important since for these first projects the terms of reference were relatively short, leaving room for different interpretations. An example of strong local ownership is project 9, which after a slow start could be implemented in a reasonably smooth way and has been finished technically (not yet formally). For this project there was relatively quickly a good contact with the local authorities, who actively supported the implementation of the project in all its aspects.

Project 16 has not advanced significantly in terms of implementation. The main reason is that in some countries the situation remained a long time unclear on whom to contact, and sometimes contact persons changed, while in e.g. Morocco the number of projects to be implemented is so high that the designated contact person had to prioritise the projects.

Project 33 aims at among others enhancing CBRN emergency response in several countries in the African Atlantic Façade and East and Central Africa. Whereas these training efforts have not yet been started, from the visits to these countries it has been noted that there exists large differences between the emergency response organisations in the various partner countries. The most extreme example is that in a certain case a fire brigade in a partner country did not always respond effectively to a fire alarm in the way a fire brigade should. In such a case one should change the objectives into efforts that focus on enhancing the classical response functions of the fire brigade rather than trying to bring these functions to the level of CBRN response, although this is formally not according to the project terms of reference.

It should be noted that this is an extreme example. In most cases CBRN response functions can be improved with available knowledge.

Participation in R&D projects is of fundamental importance to remain up-to-date with the most recent developments and to be able to train CBRN first responders on basis of these developments. SCK•CEN contributes to these developments both with R&D in more classical methods like the development of retrospective dosimetry and in the use of social media for emergency response. The development in retrospective dosimetry will enhance dose assessment of members of the public in two ways: the method is able to measure lower doses than the presently existing techniques (in the order of 100 mGy instead of presently 0.5–1 Gy) and it is much cheaper to apply, making it possible to perform an individual dose assessment of many people. The use of social media may lead to an early warning system that will support decision making of the authorities.

10.6 Conclusions

Local ownership is primordial for the success of project implementation. Local contacts should be accessible for the project implementers and should have a significant authority in the partner country in order to be able to move things.

The results of training exercises in various countries will be different in function of the existing knowledge and expertise present. Therefore the need assessments are very important in order to use the available effort in the most efficient and effective way.

Participation in R&D activities is essential; both to improve first response and to spread this knowledge among the first response community.

References

1. National Response Framework, Homeland Security, USA (2008) Retrieved online http://www.fema.gov/pdf/emergency/nrf/nrf-core.pdf
2. Federal Emergency Response Plan, Canada (2011) Retrieved online http://www.publicsafety.gc.ca/cnt/rsrcs/pblctns/mrgnc-rspns-pln/index-eng.aspx
3. National Emergency Response System, Canada (2013) Retrieved online http://www.publicsafety.gc.ca/cnt/rsrcs/pblctns/ntnl-rspns-sstm/index-eng.aspx

4. European Civil Protection (2012) EC Humanitarian aid and civil protection. http://ec.europa.eu/echo/about/index_en.htm
5. Breeding D (2014) Guide to writing standard operating procedures – Engineering Safety Office – Office of Engineering Safety – Industrial Hygiene, Texas A&M University, College Station, Texas, USA
6. Grusenmeyer D (2014) Developing effective standard operating procedures. Cornell University, College of Agricultural and Live Sciences, Department of Animal Sciences, Ithaka, NY, USA
7. Council of Europe (2004) International organisations and major hazard management – AP/CAT (2003) 12 revised. Strasbourg, Council of Europe
8. U.S. EPA (2007) Guidance for preparing Standard Operating Procedures (SOPs). United States, Environmental Protection agency – EPA QA/G-6 – EPA/600/B-07/001 – Apr 2007, Washington, DC, USA
9. Spoehr et al (2008) Multiservice tactics, techniques, and procedures for chemical, biological, radiological, and nuclear consequence management operations – FM 3-11.21 United States Army Training and Doctrine Command Fort Monroe, Virginia, April 2008
10. NATO (2008) Project on minimum standards and non-binding guidelines for first responders regarding planning, training, procedure and equipment for Chemical, Biological, Radiological and Nuclear (CBRN) incidents – guidelines for first response to a CBRN incident. NATO Civil Emergency Planning - Civil Protection Committee, Brussels, Belgium
11. IAEA (2000) Generic procedures for assessment and response during a radiological emergency – IAEA TECDOC-1162, IAEA, Vienna, Austria
12. Rojas-Palma C, Cauwels V, Woda C, Discher M (2014) Preliminary results on how CATO seeks to address public reassurance during a radiological exposure type of incident. In: 3rd international conference on preparedness and response of healthcare systems to emergencies and disasters EMS World, Tel Aviv, Israel, 12–15 January 2014

Chapter 11
EU Efforts in Managing CBRN Terror Attacks

Friedrich Steinhausler

Abstract The international security expert community foresees a growing probability for terrorist attacks using chemical (C), biological (B), radiological (R), and nuclear (N) material. Despite the high-impact such an event would have on the targeted society, the low probability of its occurrence reduces the willingness in some European Union (EU) Member States to invest in an increased level of CBRN preparedness beyond the absolute minimum. The European Commission has responded by allocating € 1,350 million for security research for the period 2007–2013, partly also for countering CBRN-related threats. This paper reports on: (1) Results of a comprehensive gap analysis among 80 EU first responder organizations (police, fire fighters, emergency medical services) in 25 EU Member States concerning the management of a mega-crisis, inter alia also concerning CBRN; (2) Analysis of major CBRN counterterrorism research achievements in the EU.

11.1 Gap Analysis

The international security community considers it feasible for terrorists to deploy weapons of mass killing (e.g., use of fully fuelled civil aircraft as guided weapons), weapons of mass disturbance (e.g., introduction of bank notes contaminated with radioactive material) or, although with a significantly lower probability, even weapons of mass destruction (e.g., deployment of crude nuclear device; [1, 2]). There is a growing probability for terrorist attacks using chemical (C), biological (B), radiological (R) or nuclear (N) material: (a) Terrorists no longer hesitate to

---

The research leading to these results has received funding from the European Community's Seventh Framework Programme (FP7/2007–2013) under grant agreements nos. 261693, 261817 and 218070.

F. Steinhausler (✉)
Government Radiological Measurements Laboratory, Salzburg, Austria
e-mail: friedrich.steinhaeusler@sbg.ac.at

S. Apikyan, D. Diamond (eds.), *Nuclear Threats and Security Challenges*,
NATO Science for Peace and Security Series B: Physics and Biophysics,
DOI 10.1007/978-94-017-9894-5_11

stage attacks leading to mass casualties; (b) Adversaries are increasingly attracted to the deployment of unconventional weapons, such as CBRN; (c) Terrorists consider attacks on nuclear installations (e.g., nuclear power plants) as desirable for their cause. Despite the high-impact such an event would have on the targeted society, the low probability of its occurrence reduces the willingness in some European Union (EU) Member States to invest in an increased level of CBRN preparedness beyond the absolute minimum. Some of the 28 Member States in the EU have gained extensive experience individually in conventional counterterrorism over decades in their fight against well organized groups. However, no CBRN-related terror attack has happened in any of the EU Member States hitherto. It required the terror attacks in New York and Washington D.C. in 2001, in Madrid in 2004 and in London 2005 to increase coordinated security research activities in EU Member States, accounting also for CBRN threats. It has to be emphasized that in several EU Member States the infrastructure of national emergency services consists to a large extent of trained volunteers, supported by full time professionals only in large cities; equipment as well as training of the former is at a lower standard. An anonymised confidential survey was conducted among 80 EU first responder (FR) organisations (police, fire fighters, Emergency Medical Services (EMS)) in 25 EU Member States on the methods currently used for managing a large-scale crisis [3]. Results showed considerable inadequacies concerning CBRN training (Figs. 11.1, 11.2, and 11.3). Within all types of EU organisations some FR trainees receive CBRN-training for a few hours only. This applies to 80 % of the EMS. On the other hand, fewer than 30 % of the firefighters and fewer than 40 % of the law enforcement organizations surveyed receive such short CBRN training only. In the EU EMS staff members are the least prepared for a response to CBRN-threats. In general, in the EU CBRN-response is viewed predominately as an issue for specialists rather than as part of general training:

- EU EMS: Basic CBRN training is provided only up to 10 h maximum; only specialists receive training exceeding 100 h.
- EU Fire Fighters: CBRN training is typically provided for up to 50 h; specialists are trained for more than 100 h.
- EU Law Enforcement Organisations: Basic CBRN training up to 50 h, but more than 100 h for specialists.

Significant deficits among EU FR were noted in the routine use of advanced IT-based tools and methods, such as:

- Application of computer-based CBRN related modelling (e.g., toxic plume dispersion; impact of conventional explosives or nuclear weapons on structures and man);
- Use of CBRN related information contained in an electronic database (e.g., concentration- and exposure limits);
- Access to decision-aiding tools concerning CBRN related risk assessment (e.g., evacuation versus sheltering).

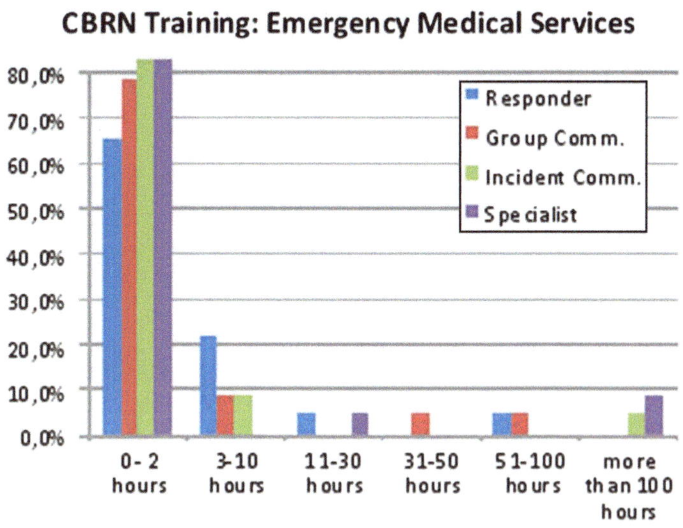

Fig. 11.1 CBRNE training for EU emergency medical services

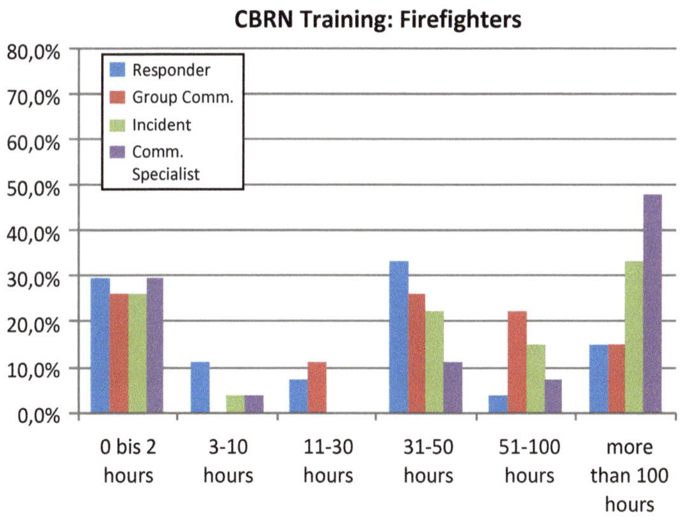

Fig. 11.2 CBRN training for EU firefighters

This frequently observable deficiency can contribute to an unnecessarily large number of victims among FR and members of the public alike in case of a CBRN incident.

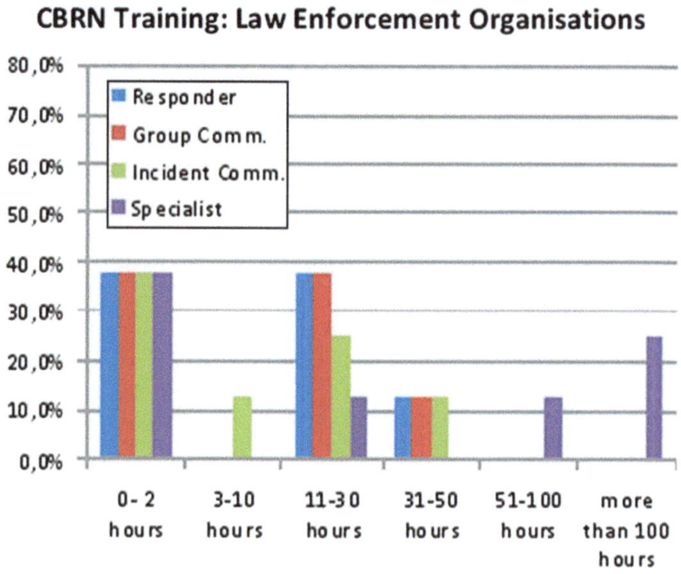

Fig. 11.3 CBRN training for EU law enforcement organisations

11.2 EU Security Initiative

The European Commission (EC) has allocated € 1,350 million for security research in its wider R&D budget – referred to as the 7th Framework Programme for Research (FP7-SECURITY) – for the period 2007–2013. The main topics are: Security of the Citizen; Security of infrastructures and utilities; Intelligent surveillance and border security; Restoring security and safety in case of crisis; Security systems integration, interconnectivity and interoperability; Security and society; Security research, coordination and structuring. This includes advanced research into the societal dimension of security, protection of citizens against man-made and natural disasters, critical infrastructure protection, crisis management capabilities, intelligent maritime and land border surveillance, pre-standardisation and the interoperability of systems. However, out of the more than 400 security-related R&D projects only 13 projects address CBRN threats [4]. Table 11.1 below summarizes the main characteristics of these EC co-funded FP7-projects. The total cost for these projects amounts to € 84,308,982.

11.2.1 Analytical Methods

In order to accelerate risk assessment and risk management it is essential to identify a CBRN agent as quickly as possible. Ability to analyze a suspected or proven

Table 11.1 FP7 security research projects related to radiological and nuclear threats

Project acronym	Full title	Coordinator	Total cost (EUR)
Analytical methods			
MIRACLE	Mobile laboratory capacity for the rapid assessment of CBRN threats located within and outside the EU	Universite Catholique de Louvain (Belgium)	1,420,617
SLAM	Standardisation of laboratory analytical methods	Uema Universitet (Sweden)	1,320,763
GIFT CBRN	Generic integrated forensic toolbox for CBRN incidents	Netherlands Forensic Institute (The Netherlands)	7,125,972
Protection of drinking water			
SECUREAU	Security and decontamination of drinking water distribution systems following a deliberate contamination	Université de Lorraine (France)	7,462,072
SAFEWATER	Innovative tools for the detection and mitigation of CBRN related contamination events of drinking water	Arttic (France)	4,814,570
Countermeasures			
COUNTERFOG	Device for large scale fog decontamination	Universidad Carlos III de Madrid (Spain)	4,407,528
IF REACT	Improved first responder ensembles against CBRN terrorism	Universite Paris XII - Val de Marne (France)	5,243,368
PRACTICE	Preparedness and resilience against CBRN terrorism using integrated concepts and equipment	Uema Universitet (Sweden)	11,691,686
CATO	CBRN crisis management: architecture, technologies and operational procedures	NESS (Israel)	14,055,519
CBRNEMAP	Road-mapping study of CBRNE demonstrator	Uema Universitet (Sweden)	1,662,022
CAST	Comparative assessment of security-centered training curricula for first responders on disaster management in the EU	Paris-Lodron Universität Salzburg (Austria)	2,858,318
Interoperability			
DESTRIERO	A decision support tool for reconstruction and recovery and for the interoperability of international relief units in case of complex crises situations, including CBRN contamination risks	E-GEOS SPA (Italy)	4,171,403
BRIDGE	Bridging resources and agencies in large-scale emergency management	SINTEF ICT (Norway)	18,075,144

hazardous agent fast and reliable is key in deciding on any further actions, whether it is protective measures for FR and members of the public, security actions or political decisions. In the ranking of priorities about 12 % of the total co-funding of CBRN FP7 projects focuses on the development of innovative analytical methods for such materials.

A rapidly CRBN deployable diagnostic and forensic capacity will be developed by 2015 which can be brought to the incident area as closely as possible inside and outside the EU (MIRACLE project).[1] It provides the innovative design of a mobile CBRN laboratory architecture, which is based on flexibility, scalability, modularity, and interoperability. The consortium addresses also structures, equipment, operational procedures, communication, logistics, forensics and related legal issues. Standardizing CBRN analysis was also at the heart of the SLAM project.[2] Besides identifying the needs and currently available procedures for CBRN sampling, transport and analysis, the consortium reviewed procedures for the "unknown sample", i.e., sample with possible content of *any* of the potential CBRN agents. The work resulted in practically applicable functional standardization of CBRN analysis. The GIFT-CBRN project[3] focuses on forensics with regard to CBRN incidents, providing procedures, sampling methods and detection of CBRN agents at the crime scene in combination with traditional forensic laboratory methods for contaminated evidence. Also, it will include laboratory methods for profiling the CBRN agents released at the incident.

11.2.2 Protection of Drinking Water

Man can live without food for up to 3 weeks, but only for about 3 days without drinking water [5]. Accordingly, the EC placed considerable emphasis on this topic. A drinking water supply is largely an open system, consisting of several catchment areas with varying vulnerability, such as raw water transfer systems, water treatment facilities, treated water reservoirs and water distribution networks. In the EU such systems frequently lack effective detection capacities, fast contamination warning systems, as well as decision support systems (DSS). About 15 % of the total CBRN co-funding by the EC is dedicated to this topic area.

Project SECUREAU[4] takes a three-prong approach: (1) Identification of sources of contamination and the contaminated areas; (2) Development of new sensors

[1]Cooperation between France, Germany, Norway, Sweden, Canada, The Netherlands, United Kingdom.

[2]Cooperation between Sweden, France, Norway, United Kingdom, Germany, The Netherlands.

[3]Cooperation between The Netherlands, United Kingdom, Finland, Belgium, Sweden, Ireland, Turkey, Spain, France.

[4]Cooperation between France, Germany, United Kingdom, Finland, Portugal, Latvia.

and identification of optimal sensor distribution; (3) Modelling of sorption and desorption, pipe wall cleaning and decontamination of sludge.

SAFEWATER addresses the key drinking water incident management challenges at large and starts from best-of-breed technologies, including an EPA challenge winning event detection system.[5] From this, the project develops a dedicated DSS for the real-time support of decision makers. This DDS offers (a) Result interpretation models to enable the real-time ranking of the severity of alerts and for the prompt identification of recovery measures; (b) Spatial detection models to determine the contamination's source and spread, using inter alia "virtual sensors", i.e., large networks of domestic sensors.

11.2.3 Countermeasures

Once a CBRN incident has occurred, speed and comprehensiveness of countermeasures taken are the determining factors for regaining control over the situation. In the EU CBRN incident management faces severe challenges: (1) Pronounced fragmentation with regard to doctrines, knowledge, processes and systems. This is reflected in significant diversity of organisational set-ups and of legacy systems for emergency preparedness and management (ICT, equipment, sensors). Acknowledging the sever deficits the EC allocated approximately 47 % of the total CBRN co-funding to improving preparedness and resilience of the EU Member States and associated countries in a CBRN terror attack.

In order to limit the spread of intentionally released CBRN material a rapid response system for collapsing all kinds of dispersed agents (smoke, fog, spores, etc.) has been developed by the project COUNTERFOG.[6] It uses a fog made of a solution which can eventually contain any kind of neutralizing component in order to (1) Neutralize and collapse the CBRN cloud; (2) Decontaminate victims and hardware in the target area. It is intended as either permanent installation in large public buildings (e.g., subway platforms, railway stations, airport terminals), or as portable system for use outdoors.

Frequently, civilian FR receive only occasional training in using their Personal Protective Equipment (PPE), experiencing heat stress and over-protection. Project IF REACT provides the next generation of PPE for civilian FR.[7] Extensive mock scenario simulations were conducted in two field exercises. Together with laboratory tests, the main shortfalls in existing training and operational PPE were identified, e.g., a pumping effect due to leaks in interfaces with gloves and boots. The new

---

[5] Cooperation between France, Germany, Israel, United Kingdom, Portugal, Sweden.

[6] Cooperation between Spain, United Kingdom, Bulgaria, Czech Republic, Sweden, Germany.

[7] Cooperation by France, United Kingdom, Germany, The Netherlands, Czech Republic, Croatia. It is noted that *nuclear threats* or *hazards* are not addressed in this project.

PPE focuses on solutions for hand- and foot protection and C4I needs of the wearer, accounting for safety, ergonomics and logistics of the civilian FR.

Two EU projects develop *CBRN Countermeasure Tool Boxes*: PRACTICE and CATO.[8,9] The approach is similar for both: (1) Identification, organization and establishment of knowledge of critical elements in the event structure by analyzing scenarios, real incidents and exercises; (2) Gap analysis in the current response situation; (3) Integration of the allocated response capabilities or functions in a toolbox of equipment, procedures and methods; (4) Development of a public information kit for decision-support, FR training and exercise. CATO adds a simulation-centered laboratory (both virtual and hosted by some of the partners).

Project CBRNEMAP uses advanced technological solutions, integrated at the system of systems level, to provide a roadmap for the subsequent construction of a CBRNE Demonstrator.[10] This prototype will be the corner stone for an optimized demonstration programme in all 28 EU Member States.

Since disaster management skills and training of the EU first responder community need to be adapted to meet the new challenges posed by CBRN material, project CAST established a comprehensive *virtual reality training* system, combining visual and acoustic impressions, motion and olfactory cognition in a 3D simulation of threat scenarios.[11] Trainee response can be evaluated using biofeed-back via wireless electrodes. Further, a curriculum for a standardized security-centered training course on disaster management was developed, recommended for all EU Member States.

11.2.4 Interoperability

Although the EC strides for harmonization of legislation and procedures, EU Member States still differ significantly with regard to CBRN related training, equipment, standards and crisis management [6]. Frequently this impedes cooperation between different FR organizations even within a given country. In order to improve interoperability between different agencies and Member States the EC allocated about 27 % of the total co-funded FP7 projects to strengthen interoperability.

The BRIDGE project, the largest CBRN-related EU project, aims for increasing the safety of citizens by developing technical and organisational solutions that

[8]Cooperation between Sweden, United Kingdom, The Netherlands, France, Norway, Czech Republic, Poland.

[9]Cooperation between Israel, United Kingdom, Denmark, Germany, Belgium, Finland, Austria, Czech Republic, Slovakia, France, The Netherlands.

[10]Cooperation between Sweden, United Kingdom, Germany, France, Italy, Belgium, Czech Republic.

[11]Cooperation between Austria, Germany, United Kingdom, Czech Republic, Spain, Hungary, Sweden.

significantly improve crisis- and emergency management in large incidents.[12] Key to this is to ensure interoperability, harmonization and cooperation among stakeholders on the technical as well as on the organisational level. For this purpose the consortium developed and tested on scene together with FR: (a) *Master System* which provides functionality to present and act on information about the incident, response and from external services; (b) *Robust and Resilient Communication* creates an ad-hoc networking infrastructure that provides networking services on an incident site to exchange data locally or send them to other networks; *HelpBeacons* application allows people to use their smart phones to advertise their need for help; (c) *Dynamic Tagging* assists first responders in marking and monitoring significant locations of the disaster site and in creating real-time situation awareness; (d) *eTriage* assists in marking and monitoring victims and in creating real-time situation awareness, thereby, it bridges the process from triage to hospital admission; (e) *Situation aWAre Resource Management* (SWARM) combines resource management (resource identification, involvement, task assignment, status reporting) with technology for achieving situation awareness; (f) *Adaptive Logistics* organizes dynamic multi-agency collaboration by using workflows to establish collaboration between various BRIDGE system components; (g) *Advanced Situation Awareness* (ASA) assists FRs on scene in increasing situational awareness by supplying real-time visual and other information on the extent of the disaster and its consequences; ASA consists of an unmanned aerial vehicle with multiple sensors, a computer-based *Expert System*, and a 3D/2D Modelling Module; (h) *Information Intelligence* allows automatic analysis of social media data in addition to live on-scene data and aggregates information in a situational report; (i) *Federated Control Room Support (FCRS) provides support for cross-agency and cross-border team formation, team process monitoring and team communication; (j) First Responders Integrated Training System* (FRITS) focuses on training, exercises and evaluation for improvements concerning training methodology tools, evaluation tools, simulated training; live, virtual and constructive systems using commercially-off-the-shelf (COTS) technology.

The next generation post-crisis assessment tool for reconstruction and recovery planning is developed in the project DESTRIERO.[13] This includes assessment of structural damage and CBRN-related contamination through advanced remote sensing. Data collection foresees the use of mobile devices by relief units on the field using novel smart-phone apps and earth observation images. Fast damage assessment and monitoring of affected areas are used in a network centric approach. Critical infrastructure recovery is given top priority, accounting for multiple social and economic aspects.

[12]Cooperation between Norway, The Netherlands, Sweden, Germany, United Kingdom, Austria, Switzerland.

[13]Cooperation between Italy, France, Spain, Germany, Poland, Ireland, Belgium, United Kingdom.

11.3 Way Forward

The topic CBRN-related terrorism is frequently a misunderstood topic among European Member States, reflected in generally underfunded activities at the national level. Although there is pronounced interest in many countries seeking training support, readiness to fund such activities is generally low as confirmed by INTERPOL for its global membership of 190 countries [7]. The EC has taken about 6 years to respond to the terror attacks in the US on 11 September 2001 before launching FP7 and co-financing a relatively modest number of CBRN related projects. The funds for the follow-up programme HORIZON 2020 (H2020) amount to € 80 billion [8]. In addition, in October 2014 the EC created a new internal users group to improve policy, legislative and operational coordination regarding large-scale disasters. This has to be seen in view of the fact that about 30 % of the 33 EC Directorates General (Commission's Departments) collaborate on topics concerning counter-CBRN activities or disaster response issues. It waits to be seen to what extent the obvious need for strengthening CBRN capabilities among EU Member States will be reflected in dedicated H2020 budget allocations and stronger links between CBRN-related operational decisions and policies by the EC and its Member States.

References

1. Steinhäusler F, Edwards F (eds) (2005) NATO and Terrorism. Catastrophic terrorism and first responders: threats and mitigation, NATO security through science series – B: physics and biophysics. Springer, Dordrecht
2. Steinhäusler F (2011) WP3 report "catastrophic terrorism", comparative assessment of security-centered training curricula for first responders on disaster management in the EU, EU FP7 CAST Project GA no. 218070 (classified)
3. Schneider B, Steinhäusler F, Zaitseva L (2011) WP1 report "report on training programmes," EU FP7 CAST project – comparative assessment of security-centered training curricula for first responders on disaster management in the EU, GA no. 218070
4. CORDIS, Community Research and Development Information Service, European Commission (2014) http://cordis.europa.eu/projects/result_en?q=contenttype=\T1\textquoteleftproject\T1\textquoteright%20AND%20\T1\textquoterightsecurity. Last visited 11 Nov 2014
5. Binns C (2012) How long can a person survive without water? Livescience http://www.livescience.com/32320-how-long-can-a-person-survive-without-water.html. Last visited 14 Oct 2014
6. Edwards F, Steinhäusler F (eds) (2007) NATO and Terrorism. On scene: new challenges for first responders and civil protection, NATO security through science series – B: physics and biophysics. Springer, Dordrecht
7. Muller J (2014) Putting on muscle. CBRNe WORLD, p 18, August 2014
8. European Commission (2014) Factsheet: horizon 2020 budget http://ec.europa.eu/research/horizon2020/pdf/press/fact_sheet_on_horizon2020_budget.pdf. Last visited 14 Oct 2014

Chapter 12
A Holistic Approach to Radiological Terrorism

Natividad Carpintero-Santamaria

Abstract The potential use by a terrorist group of a Radiological Dispersion Device (RDD) or dirty bomb is a contemplated contingency in the twenty-first century. Since the 1990s, smuggling of radioactive materials has occurred in several countries involving sufficient amounts to make a dirty bomb. The health effects of an RDD explosion may vary significantly according to particle energy, and whether alpha, beta and gamma particles are inhaled, ingested or skin-absorbed. The activity of the radioactive source is another variable as well as the shape of the terrain, population density, local wind conditions, etc. The explosion of an RDD could contaminate water supply, food and other consumer articles. The physiological and psychological damage caused by the explosion of an RDD would most likely be greater than the effects produced by the radioactive contamination. Taking into account that the perception of fear is produced through a complex process of cognitive, behavioral, emotional and psychological factors, an RDD explosion could result in dozens of individuals traumatized for each physically injured person. Experimental tests and mathematical code systems to calculate radioactive doses produced in the explosion of an RDD are being developed, together with new technologies and techniques in the preparedness for radiological response and recovery. Although a worst case scenario could be postulated, proactive and integrating response strategies would significantly reduce the damage caused by a dirty bomb explosion.

12.1 Introduction

Nuclear terrorism poses one of the most significant concerns in the struggle against terrorism. Although the acquisition of weapons-usable fissile material is a challenging task for a terrorist organization and the probability that a terrorist group could make an improvised nuclear device or crude nuclear bomb is small, a main concern comes from the potentiality that a highly motivated terrorist group, either

N. Carpintero-Santamaria (✉)
Institute of Nuclear Fusion, Polytechnic University of Madrid, Madrid, Spain
e-mail: natividad.csantamaria@upm.es

S. Apikyan, D. Diamond (eds.), *Nuclear Threats and Security Challenges*,
NATO Science for Peace and Security Series B: Physics and Biophysics,
DOI 10.1007/978-94-017-9894-5_12

acting independently or acting as part of a bigger organization, could detonate a radiological dispersion device (RDD), commonly known as dirty bomb [4, 6].

In April 2009 the Nuclear Security Summit was held in Washington. One of the main objectives of the summit was to develop an international effort to safeguard all vulnerable nuclear material worldwide in a time period of 4 years. In March 2012 the second Nuclear Security Summit was held in Seoul (South Korea). Issues of this conference such as minimizing stockpiles of nuclear materials; preventing the sabotage of nuclear reactors and establishing design basis threat, were discussed by the 53 participating countries and the 4 international organizations (United Nations (UN), European Union (EU), International Atomic Energy Agency (IAEA) and INTERPOL). At the closing press conference, South Korean President Lee Myung-bak stated: "Although the damage from radiological terrorism may not be as extensive as that of nuclear terrorism, there is a greater chance of it actually happening and the economic and psychological impact would be considerable" [21].

Although no attacks with radioactive agents have ever occurred, the relative accessibility of some of these materials and the perception of a new strategic terrorism in the twenty-first century, make us to carefully assess this threat contingency and develop strategic programs to detect and prevent it, and ultimately, to respond effectively. As Alexander [1] points out: "While a general pattern of criminal behaviour may be attributed to all terrorists, the modus operandi of terrorist groups will vary considerably depending on each group's motivation and capabilities".

12.2 Dirty Bombs or Radiological Dispersion Devices (RDDs)

The term *dirty bomb* is applied to radiological dispersion devices (RDDs) made with chemical explosives (gunpowder, dynamite, semtex, C-4) provided by a radioactive material depot (ampoule, vial or depot) that is commonly found in hospitals, industries, nuclear and biochemical research centres or food sterilization facilities.

Due to the number of nuclear illicit trafficking incidents during the 1990s, the International Atomic Energy Agency (IAEA) classified radioactive sources into five different categories from the most to least radiological toxicity. The aim of this categorization was aimed at improving risk decision and assisting the countries in their emergency preparedness plans and response to radiological incidents [12]. Category 1 refers to radioisotopes used in thermoelectrical generators for satellites (Sr-90, Pu-238, Po-210); food sterilization and hospital material (Co-60, Cs-137) and teletherapy for cancer treatment (Co-60, Cs-137). Category 2 refers to radioisotopes used in industrial gammagraphy, welding analyses and material tests (Co-60, Ir-192) and brachytherapy for cancer treatment (Co-60, Ir-192). Categories 3–5 refer to radioisotopes used in nuclear physics or biological research laboratories for tracing molecules in medical and biological research, and in positron emission tomography (C-11, F-18).

Although there are several radionuclides that could be used in an RDD, the following are specially relevant:

Table 12.1 Potential radionuclides that could be used in a dirty bomb (G. Velarde, personal communication)

Radionuclide	Half life	Particle emission
Polonium ($Po^{210}{}_{84}$)	138.4 days	**α of 5.31 MeV** and γ of 0.8 MeV
Radium ($Ra^{226}{}_{88}$)	1,600 years	**α of 4.78 MeV** and γ of 0.6 MeV
Plutonium ($Pu^{238}{}_{94}$)	86.4 years	**α of 5.50 MeV** and γ of 1.1 MeV
Americium ($Am^{241}{}_{95}$)	433 years	**α of 5.49 MeV** and γ of 0.8 MeV
Strontium ($Sr^{90}{}_{38}$)	28.1 years	**ß of 0.5 MeV**
Iodine ($I^{131}{}_{53}$)	8.0 days	**ß of 0.6 MeV** and γ of 0.6 MeV
Caesium ($Cs^{137}{}_{55}$)	30.0 years	**ß of 1.2 MeV** and γ of 0.6 MeV
Iridium ($Ir^{192}{}_{77}$)	74 days	**ß of 0.7 MeV** and γ of 0.6 MeV
Cobalt ($Co^{60}{}_{27}$)	5.3 years	ß of 0.3 MeV and **γ of 1.3 MeV**.

Table 12.1 include α, ß or γ particles that have the largest energy emitted by a specific radionuclide. Particles in bold type refer to the highest radiological toxicity. On the other hand, decay products of some of the above radionuclides can be important emitters of α, ß and γ particles. For example, Cs-137 decays to Barium 137 m (B-137 m) with a half life of 153 s. B-137 m is responsible for the emission of gamma rays of 0.6 MeV [23, 24].

Both the efficiency of the dirty bomb and its radioactive contamination level would depend on the chemical explosive used and the radiological toxicity of the material. The higher the destruction capacity of the chemical explosive, the more effective the scattering of the radioactive material will be. Nuclear forensics are being developed to analyse the attribution of radioactive materials to find out clues to their original source. This scientific area is of great importance if we take into account the number of radioactive materials that can be found in the black market trade.

12.3 Traumatic and Psychological Effects of Radiological Terrorism

The main purpose of a dirty bomb is to cause chaos, mass panic and traumatic and post-traumatic psychological-psychogenic effects in the population. In addition to these effects, terrorists might consider the economic losses and the costs of decontamination that would follow to this type of attack. The explosion of an RDD could contaminate water supply, food and other consumer articles, thus causing a great alarm and social disruption.

Although RDDs were discarded as military tactical weapons, they could be of great interest to terrorists because of the above reasons. Decontaminating the affected area could cost from a few million Euros up to several hundred million Euros, the cost depending on the category of the radioactive material [25]. During the expansion of the radioactive cloud, the radionuclides can be inhaled, ingested, deposited on the skin of people found in the cloud expansion area. In the case of explosion of a dirty bomb the number of victims, both those physiologically and psychologically injured, would create a complex and difficult task for hospital logistics.

> The treatment of casualties of radiological accidents is extremely varied and complex. They must be cared in hospitals by staff who are engaged in a daily basis in the haematological, chemotherapeutic, radiotherapeutic and surgical treatment of patients at risk from cancer, immunosupression and blood dyscrasias (The Radiological Accident in Goiania 1988).

Up to now, research made to assess the psychological effects caused by the explosion of an RDD is mainly based on experiential cases from incidents related to nuclear power plants such as Three Mile Island (1979, USA) and Chernobyl (1986, USSR). With respect to radioactivity, because of disinformation or, more important, because of the manipulated information of some sensationalist mass media, people generally have a deep fear. The accident occurred at Fukushima Daiichi nuclear power plant as a consequence of the devastating tsunami that struck Japan on March 11, 2011, put into account that social impact around the world was not derived from real facts, but from the way several mass media, antinuclear groups and irresponsible entities transmitted the events. The spreading of false news describing apocalyptic scenarios produced a great level of stress in Japanese population.

Psychological impact of a person exposed to radiation could result in severe psychological effects due to the psychosis produced by a magnified impression of the risk. If we take into account that fear perception is produced through a complex process of cognitive, behavioural, emotional and psychological factors, in the case of an RDD explosion these factors would be notoriously increased. The sarin gas terrorist attack in Tokyo subway (Japan) in 1995 can be considered as a paradigm of a terrorist attack to population with chemical warfare agents. After the attack, a great deal of the approximately 5,500 persons that went to 280 medical centres in Tokyo were citizens that, either did not suffer gas exposure harm or were not genuinely affected, but were afraid to become ill [18].

An incident involving radioactive agents happened in London when Mr. Alexander V. Litvinenko reportedly died from radioactive Polonium-210 poisoning in the intensive care unit of the University College Hospital in London on November 23, 2006. By November 26, 2006 "hundreds of people contacted the NHS [National Health Service] direct hotline to seek advice about potential radiation poisoning". "By December 4, more than 3,000 persons had phoned the British Health Protection Agency" [22]. Following the death of Alexander Litvinenko, the Health Public Agency (HPA) organized 24 monitoring teams from the Division of Radiological Protection to measure the level of exposure to Po-210 by people who may have been in places where supposedly occurred the incident and other places. According to Bailey et al. [2]:

> The alleged poisoning of Mr Alexander Litvinenko with polonium-210 was an extraordinary event that presented some unique public health challenges. Environmental polonium-210 contamination was found at a number of locations in London, including parts of two hospitals, several hotels, restaurants, and office buildings. [...] Urine samples from 753 people were processed: about 500 during the first month, another 250 up to the end of May 2007, and a further three up to August 2007

The hypochondriasis factor was also studied after the Polonium-210 incident, and research by Morgan et al. [16] put into account that "Providing reassuring messages during a public health incident may be ineffective for individuals with high health anxiety".

It is therefore of utmost importance to develop a proactive communication policy to adequately inform the population. Credibility is an essential factor to mitigate the uncertainty and alleviate the psychological impact in the potential case of an RDD explosion.

12.4 Code Systems to Calculate the Effects of an RDD Explosion

Radioactive doses produced in the explosion of an RDD have been calculated by means of experimental tests and mathematical code systems. Some of these code systems have been originally developed for potential accidents in nuclear power plants and are being implemented in several countries to calculate the effects produced by an RDD explosion. Local weather data and the activity of the radioactive source are used as initial references to calculate the inhalation, groundshine doses and cloudshine doses received by people at the area surrounding the explosion. Some of these code systems are:

- HOTSPOT, developed by the US Lawrence Livermore National Laboratory, [11].
- HPAC (Hazard Prediction and Assessment Capability), developed by the US Defense Threat Reduction Agency [9, 10].
- RODOS (JRODOS) (Real-time On-line DecisiOn Support), developed under the auspices of the Framework Programs of the European Commission [19].
- LASAIR (Lagragian Simulation of Dispersion and Inhalation of Radionuclides), conducted by the German Federal Office for Radiation Protection [26].

The three first codes are based on the Gaussian algorithm whereas the LASAIR code uses the Lagrange algorithm and is used to follow up on local atmospheric variations after an explosion. The prototype version of the RODOS code includes both algorithms. If we compare these code systems under the same initial conditions analogous results are obtained from a downwind distance of about 700 m. At inferior downwind distances, the RODOS code system obtains lower inhalation doses.

HOTSPOT is a model that provides fast results and that is easy to use. It is not applicable in areas with complex topography and weather. The data input is limited to a fixed windspeed and direction and atmospheric stability in fields, which vary from 0.1 to 1.0 m. Output data are limited to inhalation doses because they are several orders of magnitude superior to the groundshine doses. HOTSPOT considers that only 10 % of the radioactive material from an RDD is dispersed as particles with a diameter of less than 1 μm that could produce lethal effects with a bigger probability.

HPAC considers complex topography and simulates building structures. This code system also includes all radioactive decay and daughter isotopes of the radioactive materials produced by the RDD from the exact moment of the explosion. When comparing the HPAC and HOTSPOT models for a similar scenario, apart from their own characteristics, HOTSPOT needs 15–30 s to obtain the doses, and HPAC requires 1–2 min.

The RODOS code system is flexible and can transmit the obtained results in real-time to European Union countries. This system uses weather forecast data provided in real-time by regional and national weather services as inputs. The inputs also include statistical data on the population and agricultural and cattle production. Hydrological data from rivers and lakes are also included. The output consists of radionuclide deposition up to distances of approximately 80 km, individual doses from external and internal exposure, and countermeasures such as sheltering, evacuation, relocation, decontamination and food and water control. Administering iodine tablets to counteract I-131 in the thyroid gland is also another countermeasure taken into account. The RODOS prototype version, PRTY-40, combines Eulerian and Lagrangian algorithms and includes abrupt terrains and meteorological dispersion in its input data. The most characteristic output data from RODOS are the calculation of individual radiological doses and the countermeasures that must be applied. These output data and countermeasures can be transmitted in real-time to European Union countries [20]. The above code systems are implemented in a regular time basis.

The IXP (International Exchange Program) allows access to codes of the National Atmospheric Release Advisory Centre (NARAC) at Lawrence Livermore National Laboratory (LLNL).

Decision Support Systems (DSS) such as ARGOS, DIADEM, CATO, NERIS-TP, etc. are developing technologies and techniques in the preparedness for CBRN emergency response and recovery.

12.5 Uncontrolled Radioactive Material: Illicit Trafficking, Smuggling and Orphan Sources

Among nuclear materials, we must distinguish between fissile materials (materials that undergo fission with neutrons of any energy) such as U-235, Pu-239, and Pu-241, and fissionable materials (materials that undergo fission with neutrons of energy superior to the threshold) such as U-238. Fissile materials are the foundational materials for atom bombs and for the primary process of thermonuclear bombs. Many other radioactive materials are used in several fields in industry, medicine, biology, hydric resources, energy production, agriculture and research.

The International Atomic Energy Agency (IAEA) established an Illicit Trafficking Database (ITDB) with information provided by state members about the illegal acquisition, use, possession and trafficking of radioactive and nuclear materials. This database reports incidents involving illegal trade and trafficking across borders, and incidents involving losses and the discovery of uncontrolled radioactive sources.

During the period 1993–2014, a total of 2,477 confirmed incidents were reported to the IAEA by participating and some non-participating states. Among these incidents, 16 confirmed incidents involved unauthorized possession of high enriched uranium (HEU) and high enriched plutonium (HEP) that were acquired in illegal transactions in international borders [13]. However, the ITDB does not contain the total number of actual incidents. This discrepancy results from various factors, including the fact that several countries are not IAEA party members and that some governments are reluctant to give a complete account of the cases, either because of national security concerns or because they do not like to reveal vulnerabilities in their security programs.

Law enforcement and intelligence agencies also have their own databases. In the case of Interpol, the Project Geiger collects own law enforcement data on illicit trafficking and information provided by other organizations and sources. Also Interpol's Operation Fail Safe acts as a counter nuclear trafficking initiative. Other institutions have their own reporting databases for nuclear and radioactive incidents. The Theft and Diversion Incident Analysis System (THADIAS) is carried out by the US Argonne National Laboratory [14]. Also in the United States, several centres have their own measures for reporting illegal nuclear and radioactive trafficking such as the NIS Nuclear Trafficking Abstracts Database (James Martin Centre for Non-Proliferation Studies, CNS) or the Database on Nuclear Smuggling Theft, and Orphan Radiation Sources (DSTO, Salzsburg University and Stanford University).

On June 24, 2002, the IAEA reported that there was an inadequate control of a number of radioactive sources that could be used to make dirty bombs. The report noted that over 100 countries were not applying sufficient security regulations to avoid the theft of what are known as *orphan sources*. Orphan sources are uncontrolled sealed sources of radioactive material in irregular conditions, and they may be acquired as result of a lack of safety or security, a lack of licensing, theft, loss or illegal possession, etc.

There is a particular concern about radioactive materials in Russian Federation territory related to the security for radioisotope thermoelectric generators (RTGs) that were abandoned during the dissolution of the Soviet Union. A RTG is a thermoelectric generator powered by the radioactive decay of different radioisotopes, such as Cesium 137 (Cs-137) or Strontium 90 (Sr-90) which are both beta emitters, and Polonium 210 (Po-210) which is an alpha emitter mainly used in military space satellites. These devices have been manufactured by the former USSR for use in unmanned automatic systems as power sources for electricity, weather stations, navigation beacons and lighthouses in remote zones. In the case of the former Soviet territory, RTGs were mainly located along the Northern Sea Route, the Arctic Coast and the Far East, but other republics also had a number of these devices.

After the dissolution of the USSR, Russian authorities intended to establish control over Soviet RTGs. According to Grigoriev [7]:

> 1007 (the number is being updated now) Strontium-90 RTGs in total have been manufactured to serve as surface power source. [...] In May 2011, there were 224 RTGs in operation. 217 RTGs in the storage facilities. 3 RTGs were lost in the preceding years. 30 RTGs are being disassembled or waiting for disassembly at Rosatom's enterprises.

Apart from the security concerns, the RTGs could also pose a local environmental hazard because they are the object of theft or vandalism. These thefts are mainly perpetrated by scrap metal collectors who dismantle the stainless steel, aluminium or lead used to shield the radioactive source and then, probably unaware of the danger, abandon the radioactive source. To address this risk, the Russian Federation government launched in 2002 the National Plan of Action for Marine Environment Protection from Local Environment Hazard Anthropogenic Pollution in the Arctic Region of Russia. Other projects are being carried out by the All Russian Scientific Research Institute of Technical Physics and Automatization (VNIITFA). The Norwegian government is collaborating with the Russian Federation to recover lost RTGs through the Norwegian Plan of Action for Nuclear Safety and the Environment [17]. In 2004 the US Department of Energy (DOE) launched the Global Threat Reduction Initiative (GTRI) to upgrade security measures to be applied worldwide at sites with inadequate protection of radioactive materials. The GTRI is cooperating with the Russian government to improve methods for recovering lost Soviet RPGs.

The European Union is developing technical assistance programs for the prevention of illicit trafficking for former Soviet republics through Technical Aid to the Commonwealth of Independent States (TACIS), the International Nuclear Society Council (INSC), and EU Instrument for Stability programs.

12.6 Detection of Illegal Radioactive Material and International Cooperation

The control and surveillance of illegal trafficking, smuggling and underground markets for weapons, narcotics, immigrants, and so forth is a huge task which becomes more difficult when manual and visual inspections are subject to legal grounds and non-invasive inspections. About ten million shipments of radioactive materials are transported around the world annually, being 95 % of these shipments small amounts of substances for medical diagnosis and treatment, agriculture or advanced research. According to authorities, mobile radioactive sources can be lost or stolen while they are in transit. However, and taking into account that detection of illegal radioactive material must become a priority for security, scientific and technological research is being conducted in several countries, and advances are increasingly improving the capabilities of radioactive detection.

Present radionuclide identification devices (RIDs) are provided with on-board ROI (regions of interest or potential targets) and are able to calculate the radionuclide activity and detect and identify large series of radioisotopes within seconds. These RIDs also provide an increasingly reliable method for showing the percentage presence of different radionuclides. Other scientific approaches include numerical simulations and active technology by detecting the induced radiation in materials. "The photonuclear technology seems very promising for detection of nuclear materials. It provides high sensitivity, wide range of detectable materials, accuracy of localization, low activation of inspected object" [3].

In 2008 the European Union Council established the New Lines for Action in Combating the Proliferation of Weapons of Mass Destruction and their Delivery Systems. Some of the main objectives of this strategy are to take measures to combat intangible transfers of knowledge and know-how, and to intensify efforts to combat proliferation financing, among others.

Spain has signed a bilateral cooperation agreement with the United States and now participates in the Megaports Initiative. According to port authorities, from January to October 2014, 5.093.814 standard containers (TEUs, 20-ft equivalent units), passed in transit through the major Spanish ports of Valencia, Bahía de Algeciras, Barcelona, Las Palmas, and Bilbao [15]. These ports are among the first hundred ports worldwide and among the top 21 European ports in volume of container traffic.

Preventing illicit trafficking in radioactive materials is of paramount importance due to the possibility that they could be transported by people who might enter illegally into Western countries. According to the European Agency for the Management of Operational Cooperation at the External Borders of the Member States of the European Union (Frontex), in the second quarter of 2013, 24,805 illegal entries and 118 clandestine entries were detected at EU border crossings [8].

12.7 Conclusions

Terrorism continues posing a threat for both national and international security and there is a general consensus that its eradication is still an arduous matter that cannot be foreseen in a next future. Related to radiological terrorism, we can conclude the following:

- Although the threat of nuclear terrorism is more complex, the threat of radioactive terrorism presents growing challenges which must lead to the reinforcing of its prevention. In a non-negligible number of illicit trafficking incidents, sufficient amounts of radioactive materials to make a dirty bomb were involved. In several cases the reasons for this smuggling were purely commercial, but in other cases the reasons have not been clearly identified.
- The probability that a terrorist group could make a dirty bomb is high, but the lethality could be reduced with an adequate emergency response. Apart from the acute and subsequent health effects, the explosion of a dirty bomb would cause social, psychogenic and psychological effects that would hinder overcoming the difficulties and circumstances caused by its potential explosion. It is likely that the psychological effects produced by the mass panic and social chaos would be greater than the effects of radioactive contamination.
- Rigorous control of radioactive sources and radioactive wastes in hospitals, industries, laboratories, and so forth must be applied in all countries. There are still vulnerabilities in cross-border trafficking that necessitate reinforcing policies to efficiently prevent and detect the radioactive threat. Multilateral and regional cooperation is essential to safeguard radioactive materials shipments.

- Regarding counterterrorism, measures such as physically securing radioactive materials, implementing technical methods for radionuclide identification and providing assistance programs to countries that lack the necessary technical capabilities are among the main multifaceted counterterrorist strategies. In certain countries there is the risk of negligence when dealing with radioactive materials. Another potential security risk is also the possibility that insiders may collaborate to clandestinely divert radioactive materials. Insiders may act under ideological, economical, sentimental or doctrinal affinities or motivations.
- It is necessary to reinforce measures to efficiently control the import and export of radioactive materials and dual-use technological equipments. Moreover, supplying radioactive materials to countries that may lack adequate safety or security controls in particular should be controlled.

The threat of extreme and violent terrorist practice is a challenge that transcends national borders and is presently one of the most disruptive asymmetric threats to international security.

References

1. Alexander Y (ed) (2002) Combating terrorism. Strategies of ten countries. University of Michigan Press, Ann Arbor
2. Bailey MR, Birchall A, Etherington G, Graser G, Wilkins BT, Bessa Y, Bishop L, Brown J, Dorrian MD, Ewers LW, Fell TP, Ham GJ, Hammond DJ, Forrester S, Hodgson A, Howarth C, Ibrahimi ZF, Maguire H, Marsch J, Phipps AW, Puncher M, Ruggles R, Shutt A, Smith JRH, Turbitt D, Youngman MJ, Wilding D (2008). Individual monitoring conducted by the Health Protection Agency in the London Polonium-210 Incident. Health Protection Agency. HPA-RPD-067. Retrieved October 16th, 2014 from https://www.gov.uk/government/uploads/system/uploads/attachment_data/file/337140/HPA-RPD-067_for_website.pdf
3. Bakalyarov A, Karetnikov M, Lebedev V, Muradyan G, Skorik Y, Tupikin N, Yakovlev G (2008) Experimental model on the device for the detection of nuclear materials by pulsed photoneutron technology. In: Apikyan S, Diamond D, Way R (eds) Prevention, detection and response to nuclear and radiological threats. The NATO science for peace and security series – B. Physics and biophysics. Springer, Dordrecht, pp 173–181
4. Carpintero-Santamaría N (2012) The incidence of illegal nuclear trafficking in proliferation and international security. Behav Sci Terrorism Polit Aggress 4(2):99–109, May 2012
5. Carpintero-Santamaría N (2012) Is WMD terrorism a likely prospect in the future? YES: WMD terrorism: a potential threat to international security. In: Jackson R, Sinclair SJ (eds) Contemporary debates on terrorism. Routledge, London, pp 76–83
6. Carpintero-Santamaría N (2014) Factors affecting nuclear security. In: Martín Ramirez J, Morrison C, Kendall A (eds) Conflict, violence, terrorism and their prevention, Cambridge Scholars Publishing, pp 150–163
7. Grigoriev AS (2011) Urgency of RTG decommissioning work management and main results of international cooperation in this sphere. Retrieved 27 March 2012 from http://www.iaea.org/OurWork/ST/NE/NEFW/Technical_Areas/WTS/CEG/documents/25th-IAEA-CEG-Plenary-Meeting/2.5-Grigoriev-Eng.pdf
8. Frontex Fran Quarterly. Quarterly, October-December 2013. http://frontex.europa.eu/assets/Publications/Risk_Analysis/Annual_Risk_Analysis_2013.pdf

9. HPAC general and specific characteristics for model. Retrieved 1 September 2011 from www.ofcm.gov/atd_dir/pdf/hpac/pdf
10. Hill A (2003) Using the Hazard Prediction and Assessment Capability (HPAC) hazard assessment program for radiological scenarios relevant to the Australian Defence Force, DSTO-CR-0294
11. Hoffman SG (2010) HotSpoT health physics codes. Version 2.07.1 user's guide. Lawrence Livermore National Laboratory. LLNL-TM-411345 Rev. 1. 8 March 2010. https://narac.llnl.gov/operational.html
12. IAEA Safety Standards (2005) Categorization of radioactive sources. N° RS-G-1.9. Retrieved on 12 October 2014 from http://www-pub.iaea.org/MTCD/publications/PDF/Pub1227_web.pdf
13. IAEA Illicit Trafficking Database (ITDB). Retrieved 10 September, 2014 http://www-ns.iaea.org/downloads/security/itdb-fact-sheet.pdf
14. Lindley RA, Adduci JJ, Johnson L, Lepoire DJ (2008) Perspectives on international radiological trafficking. In: Apikyan S, Diamond D, Way R (eds) Prevention, detection and response to nuclear and radiological threats. The NATO science for peace and security series – B. Physics and biophysics. Springer, Dordrecht, pp 47–55
15. Ministerio de Fomento (2014) Retrieved on 17 October 2014 from http://www.puertos.es/estadisticas/estadistica_mensual/index.html
16. Morgan OW, Page L, Forrester S, Maguire H (2008) Polonium-210 poisoning in London: hypochondriasis and public health. Prehospital Dis Med 23(1):96–97, Retrieved on March 25, 2012 from http://pdm.medicine.wisc.edu/Volume_23/issue_1/morgan.pdf
17. NRPA Bulletin. Stralevern Info (2004). Norwegian radiation protection authority. www.stralevernet.no. Retrieved 25 July 2011
18. Pangi R (2008) Consequence management in the 1995 sarin attacks on the Japanese subway system. In: Howard RD, Forest J (eds) Weapons of mass destruction and terrorism. McGraw Hill, Chicago
19. Raskov W (2008) The Real-time On-line Decision Support System RODOS. Preparedness and response for nuclear and radiological emergencies. Retrieved on 3 September 2011 from www.iscram.org/dmdocuments/P269_278.pdf
20. RODOS Homepage (2008). Retrieved on 3 September 2014 from http://www.rodos.fzk.dehttp://www.rodos.fzk.de/
21. 2012 Seoul nuclear security summit. Beyond security towards peace. Retrieved on 27 March 2012 from http://www.thenuclearsecuritysummit.org/eng_media/press/press_view.jsp?oCmd=6&b_code=1&idx=302&rnum=1&f_gubun=0
22. Timeline: Litvinenko death case. Retrieved 25 March 2012 from http://www.news.bbc.co.uk/2/hi/6179074.stm
23. Velarde G (2008) Nuclear terrorism: an overview. In: Apikyan S, Diamond D, Way R Prevention, detection and response to nuclear and radiological threats. The NATO science for peace and security series – B. Physics and biophysics. Springer, Dordrecht, pp 11–17
24. Velarde G (2014) Personal communication
25. Velarde G, Carpintero-Santamaría N (2010) Global terrorism: an assessment of biological, chemical and nuclear threat. In: Antonius D, Brown AD, Walters TK, Martín Ramirez J, Sinclair SJ Interdisciplinary analyses of terrorism and political aggression, Cambridge Scholars Publishing, pp 239–271
26. Walter H (2001). Dose assessment after an explosive dispersion of radioactive material in urban surroundings. A decision support system for use by radiation protection authorities. In: Proceedings of the 7th international conference on harmonisation within atmospheric dispersion modelling for regulatory purposes, Belgirate, 28–31 May 2001

Chapter 13
The Strategic Impact of an Iranian Nuclear Weapons Capability on Israel

Nadav Morag

Abstract This paper will address the likely strategic impact of an Iranian nuclear weapons capability on Israeli security, both in terms of the country's regional standing within the Middle East, and in terms of its homeland security issues. It should be emphasized that an Iranian capacity to produce and deploy nuclear weapons in a fairly short period of time will have largely the same strategic impact on Israel as an already existing Iranian nuclear weapons capability because Iran will be able to claim that, by developing this capacity, it will be able to counter Israeli "aggression" in the Middle East, thus enhancing its prestige in the region and beyond. Moreover, an Iranian capability to develop and deploy nuclear weapons may embolden Iran to risk further confrontation with Israel, the United States, and America's Arab allies because a nuclear weapons capability is likely to be perceived in Teheran, particularly by regime hardliners, as an insurance policy against a catastrophic attack on Iran that could threaten the regime's hold on power. Finally, even if Iran does not actually build nuclear weapons, once it has the capacity to build them in a short period of time, Israel will need to think about the implications of their use against Israeli cities and what this means for its homeland security.

Given the aforementioned, Iran does not need to actually possess nuclear weapons in order to reap the benefits of being a nuclear power, particularly since there is little evidence to suggest that Iran, although an extremist power, acts irrationally and would actually want to use such weapons against Israel. Despite the anti-Israeli views held by Iran's leadership, Tehran's relationship with Jerusalem seems to have been impacted by Iranian perceptions of their geopolitical interests rather than motivated solely, or even largely, by ideology (Trita Parsi, *Treacherous Alliance: The Secret Dealings of Israel, Iran, and the United States* (New Haven: Yale University Press, 2008), Kindle Edition, location 3581.). The problem for Israel is that Iran views itself as a major regional player, given its size and historical importance, and this means that it will almost necessarily view Israel, which is arguably the strongest military power in the Middle East, as a rival. This tendency is reinforced by the understanding in Teheran that the key

N. Morag (✉)
Center for Homeland Defense and Security, Naval Postgraduate School, Monterey, CA, USA
e-mail: nmorag@nps.edu

S. Apikyan, D. Diamond (eds.), *Nuclear Threats and Security Challenges*,
NATO Science for Peace and Security Series B: Physics and Biophysics,
DOI 10.1007/978-94-017-9894-5_13

to enhancing Iran's regional influence and undermining American power lays in gaining Arab allies rather than in developing friendly relations with Israel (Dalia Dassa Kaye, Alireza Nader, Parisa Roshan, *Israel and Iran: A Dangerous Rivalry* (Santa Monica: RAND, 2011), p. 56.). Moreover, Iran takes its self-proclaimed role as a leader in the Islamic world seriously and this too, given the widely perceived view that Israel is occupying the city of Jerusalem, with its Muslim shrines, and oppressing the (mainly Muslim) Palestinian people, also ensures that it stands to benefit more from a state of rivalry with Israel than one of rapprochement (Michael Eisenstadt, *The Strategic Culture of the Islamic Republic of Iran: Operational and Policy Implications*, MES Monographs, No. 1 (Quantico, VA: Marine Corps University, 2011), p. 3.). This is not to suggest that Iran will have an interest in risking its own security by engaging in a nuclear war with Israel, only that having a nuclear capability of some sort (either existing or easily acquired when needed) will provide Iran with more leverage in its "Cold War" with Israel.

This analysis will be organized around three elements: (1) the nature of recent changes in the Middle East and their implications for the rivalry between Israel and Iran, (2) how an Iranian nuclear capability (realized or potential) may impact that rivalry and events in the region more broadly, and (3) a brief discussion of Israeli homeland security/civil defense issues in the context of the extremely low probability, but unimaginably high consequence, possibility of an Iranian nuclear strike on Israel.

13.1 The Radically Changing Middle East: Implications for the Rivalry Between Iran and Israel

While most commentators could not predict it at the time, the Arab uprisings in early 2011 have led to the breakup of Libya, Syria, and Iraq and are currently threatening the stability, such as it is, of Lebanon. Egypt, arguably the most important Arab country, is holding together with generous amounts of financial aid from some of the Persian Gulf states, but has had to deal with an insurgency of sorts in the Sinai and is absorbed in suppressing the Muslim Brotherhood. Both Jordan and Saudi Arabia are threatened by the rise of the Islamic State (IS) in Syria and Iraq and there is no guarantee that these countries will remain stable in the long term.

For the last several decades, particularly since the Iranian revolution of 1979, Middle East politics has been characterized by the push and pull of relations between Israel, Iran, and the major Arab countries. At present however, the intense disarray in which the Arab World finds itself, with key countries such as Syria and Iraq unable to play their traditional role, means that there are fewer major actors to compete for regional influence. Essentially, the list of the main actors in the region can be limited, at present, to Egypt, Israel, Saudi Arabia, and Iran, with Turkey and some of the Gulf States playing a lesser role. As Israel's Operation Protective Edge and the subsequent ceasefire between Israel and Hamas have shown, diplomatic cooperation between Egypt and Israel is extremely close (albeit largely behind the scenes) and it is likely that the Saudis (Egypt's primary financial supporter) are closely involved. This leaves Iran severely isolated in the region. While it still has allies and surrogates in the Middle East, most notably Hezbollah and the Assad regime in Syria, the political changes in Baghdad could result in diminished Iranian influence over Iraq—either due to the efforts of a more inclusive government under the new Prime Minister, Haider al-Abadi, to woo Sunnis and Kurds (which necessarily requires distancing from Iran), or under the scenario that the Iraqi state collapses completely (in which case Iran would only have influence over the rump Shi'a part of Iraq). In other words, the Syrian civil war, the overthrow of the Muslim Brotherhood government in Egypt, and the rise of IS, have presented Iran with an acute challenge and have resulted in a significantly diminished regional influence. Given this situation, it is likely that Iran will try to attempt to regain some of that influence through stabilizing Iraq (and ensuring its new government takes into account Iranian interests through ensuring continued Shi'a control[1]), supporting Assad's effort to regain control of Syria's largest city, Aleppo (as part of his attempts to claw his way back to a position of control over significant portions of Syria), and trying to coax Saudi Arabia and Egypt away from their current relationship and commonality of interests with Israel. This should be possible because—while Egypt and Saudi Arabia are currently focused on what they perceive as the "Muslim Brotherhood threat" in, among other things, its Hamas manifestation—the growing power of IS, which has deployed to the Saudi Border with Iraq, means that Iran can make a case for common cause with the Saudis and their Egyptian allies against this new and frightening Jihadist threat. While criticism of Iran in Sunni Arab countries has been fierce, owing to the Islamic Republic's support for the Assad regime in Syria and the general suspicion of "Persian intentions" in the region, Sunni Arab states such as Saudi Arabia and Jordan are becoming increasingly fearful of the growing power of radical Sunni Jihadist groups such as IS. Given that Iran, and its clients, the Syrian and Iraqi governments, are probably in the best position to blunt and eventually push back the advances of IS and Al-Qaeda affiliates such as al-Nusra, there are grounds for predicting that a rapprochement of sorts will emerge between Iran and moderate Sunni Arab countries. As Israel has little to offer the Saudis and others in the moderate Sunni camp with respect to the fear of growing Jihadi strength in Syria and Iraq, they are more likely to be interested in cooperation with Iran, provided it undertakes not to try and influence domestic politics in Arab countries, and this will ultimately strengthen Iran's hand.[2]

The seeming potential commonality of interests between Iran and moderate Sunni Arab states due to the IS threat also appears to hold for relations between Teheran and Washington. Both sides have a strong common interest in weakening IS and other Jihadi groups and in creating a stable Iraqi government—though, with

[1]Hayden and Evan [1], p. A11.

[2]"Saudi Arabia Sets Condition for Iran Cooperation," *Al Arabiya News*, August 29, 2014. Available at: http://english.alarabiya.net/en/News/middle-east/2014/08/29/Saudi-Arabia-sets-condition-for-cooperation-with-Iran-.html (accessed, September 2, 2014).

the caveat that each side interprets stability in Iraq differently and, for the Iranians, stability means that Iraq's government must continue to maintain its "special relationship" with Teheran.[3] These, at least partially-shared interests, coupled with the anemic relations between Jerusalem and the Administration in Washington, make a rapprochement of sorts between Washington and Teheran more likely. If a nuclear deal is signed (a deal that could still leave Iran in possession of the capacity to manufacture nuclear weapons), this could lead to stronger US-Iran relations without the need for Iran to soften its approach towards Israel, particularly if it perceives American-Israeli relations as having been downgraded due to public spats between the Netanyahu Government and the Obama Administration. Consequently, here too, there is little hope for an Iranian assessment that it will be in their interests to reduce their hostility towards Israel.

Confronting Israel also provides potential benefits for Iran in trying to delegitimize its Jihadi rivals in Syria and Iraq. While moderate Sunni Arab states have been attacked by Jihadi groups for failing to confront Israel, a more hostile Iranian approach to Israel could allow the Iranians to argue that they are resisting "Israeli aggression" while the Jihadi groups only provide lip service to this idea and instead focus on killing other Muslims. Indeed, Iran's ally, Hezbollah Secretary-General Hassan Nasrallah, has claimed on at least one occasion that the Jihadis are furthering Israeli interests.[4]

13.2 Closer to Home: Iran's Impact on Hezbollah and the Palestinian Organizations

A confrontational approach towards Israel on the part of Iran could also help it rehabilitate Hezbollah, an organization who's legitimacy in the eyes of Sunni Arabs was significantly degraded after that organization seemingly decided to abandon the path of resistance to Israel in favor of propping up the hated Assad regime. A shift in Hezbollah's focus back to Israel could help it try and rebuild some of its lost legitimacy, and Iran's influence over Hezbollah could be exercised towards this end. However, such a change is likely to occur only gradually because, with Hezbollah's recently announced redeployment of at least some of its forces from Syria to Lebanon, it appears that the Lebanese Shi'a organization is gravely concerned with the threat that IS and other Sunni Jihadi groups pose to its position in Lebanon.[5]

Finally, in the wake of the 50 days of fighting between Israel and Hamas in the context of Operation Defensive Edge, Iran has the opportunity to rebuild its

[3]Hashem [2].

[4]Frantzman [3].

[5]"Report: Hezbollah Units Returning to Lebanon From Syria," *The Algemeiner*, August 27, 2014. Available at: http://www.algemeiner.com/2014/08/27/report-hezbollah-units-returning-to-lebanon-from-syria/ (accessed September 2, 2014).

relationship with Hamas, which was seriously damaged when Hamas decided to side with the rest of the Sunni Arab world against the Assad regime in Syria. Given the damage sustained to Hamas's infrastructure during Operation Protective Edge, Hamas will be looking to resupply itself with contraband and Iran can use this to rehabilitate this relationship and use it to gain credibility in the Arab World and as another tool with which to be seen to confront Israel. Nevertheless, there is likely to be continuing distrust on Iran's part of Hamas given that the organization abandoned Iran's Syrian ally and it is heavily dependent on Qatar (and to a much lesser degree, Turkey), and these countries are not interested in Iran enjoying a greater degree of influence in the region.

Iran's continued pursuit of nuclear enrichment, nuclear weapons technology, and delivery systems, even if they do not lead to the actual production of a bomb (just the potential for comparatively rapid production and deployment), can serve to enhance Iran's position on all these fronts for three reasons. Firstly, it will greatly enhance Iran's prestige as the country that was able to achieve a nuclear capability despite Israeli military threats and Western economic sanctions. The Iranians can thus portray themselves as having successfully stood up to Israel and the West in order to realize their "national rights"—something that will gain regional admiration in the same way that Saddam Hussein was lionized by many in the region for his willingness to face the United States prior to the first Gulf War. Secondly, Iran can portray this nuclear capability, or capacity, as an effective way of deterring Israel and thus containing what many view in the region as Israel's tendency to try and dominate the Middle East (while this is completely at odds with Israel's view of its own role, it is nonetheless the popular perception). Thirdly, as a *de jure* or de facto nuclear power, Iran can claim that this capacity will deter American or other Western intervention in the region and thus that Iran can ostensibly defend the region from Western encroachment.

13.3 Potential Israeli Responses

Israel, for its part, has few options to try and deter and/or contain Iran. Jerusalem is likely to continue to strengthen its relationship with the Kurds, particularly those in Iraqi Kurdistan. An independent Kurdish state would likely lead to the permanent partition of Iraq into sectarian states, something that would weaken Iran's position since it would leave Iran influential only in the Shi'a rump of the Iraqi state and would cut the geographic link between Iran, Iraq, Syria, and Hezbollah in Lebanon (the so-called "Shi'a Crescent" referred to by King Abdullah of Jordan in 2004). Therefore, such a development is likely to be in Israel's interest (aside from the fact that it provides Israel with a regional ally in an otherwise generally hostile geographic region, particularly given the present Turkish government's official attitude towards Israel).

Over the last several years, Israel has attempted to cajole Western states to employ sanctions in order to slow, and, if possible, halt, Iranian nuclear development.

While it is far from clear that the P5+1 countries (the US, Russia, China, France, the UK, plus Germany) and Iran will reach an accommodation regarding Iran's nuclear development, reports suggest that such an accommodation would leave Iran with the capacity to replenish the stocks of highly enriched uranium (HEU) which it would agree to destroy as part of a nuclear deal. The key elements of a potential agreement include: an Iranian commitment not to enrich uranium above 5 %, a commitment to dilute existing stockpiles of near 20 % enrichment to the 5 % level, a commitment not to install additional or more advanced centrifuges and not to use most of the centrifuges for enrichment, and a commitment not to commission the nuclear reactor at Arak (which could be used to produce plutonium).[6] The signing of a deal, however, would still not prevent Iran from achieving nuclear capability (whether realized or potential) as estimates suggest that existing older Iranian centrifuges at the Natanz site, which would be kept in place under the nuclear deal, could produce enough Highly Enriched Uranium (HEU)—approximately 25 kg—to produce a nuclear bomb in 6 months.[7] In fact, an agreement between Iran and the P5+1 would further delegitimize (in terms of international law, international public opinion, and the political support of Israel's allies) an Israeli military operation against Iran's nuclear facilities, thus possibly provide additional ammunition to those in the Israeli intelligence and policymaking communities who are arguing against such an attack.

Of course, nuclear deal or no nuclear deal, Israel could still attack Iran as a last resort, and attempt to destroy Iran's key nuclear facilities—presumably including its enrichment facilities at Esfahan, Natanz, and Fordow, and the reactor at Arak.[8] However, it is not at all clear that such an attack would succeed in significantly delaying Iran's ability to develop nuclear weapons.[9] An American strike may meet with more success, owing to the United States' ability to carry out extended operations over great geographic distances, something that Israel has difficulty doing. But here too it seems unlikely that the United States would authorize such an attack. After all, the US chose not to respond to North Korea's development of nuclear weapons through military means even though one could argue that North Korean nuclear weapons represent as much of a threat to American interests as Iranian ones. Ultimately, the United States may simply not view a nuclear-capable Iran as a fundamental threat to its national interests, and certainly not an existential threat, in the way that some, though not all, Israelis do. In the final analysis, it is certainly feasible to assume that Israel will not undertake military operations against Iran, particularly if Iran eventually reaches an agreement with the P5+1 countries.

[6]The White House [4].

[7]Kimball [5].

[8]Zanotti et al. [6], p. 37.

[9]Ibid., p. 41.

13.4 Defense and Deterrence: Israeli Response to an Operational Iranian Nuclear Capability

While it seems unlikely that Israel will be able to ultimately prevent Iran from developing nuclear weapons should it choose to do so, Israel still has a number of options for deterring any Iranian attempt to attack Israel with nuclear weapons. While it seems unlikely that the ruling regime in Iran is irrational, and therefore unlikely to ever be prepared to use nuclear weapons against Israel, Israel's development, with American funding, of the Arrow anti-ballistic missile defense system means that Israel has the likely ability to intercept some or all Iranian nuclear missiles fired at Israel. Of course, no one can say whether the Arrow will prove successful in an operational setting in the way that it has in tests conducted in 2009 and 2011.[10] Success in this context is dependent on the radar, tracking, and interceptor systems working perfectly as well as having a limited number of targets, but it could be sufficient, coupled with an Israeli second strike nuclear capability, to deter Iran from any thoughts of nuclear war with Israel. Israel reportedly has a nuclear cruise missile capability on its three German-made Dolphin class submarines (a fourth submarine has just been released to Israel) as well as some 200 nuclear-capable Jericho missiles.[11] It is likely that in order for deterrence to be effective, Israel will need to do away with its heretofore opaque nuclear policy and become an overt nuclear power.[12] Needless to say, both active defenses and a second-strike capability can only effectively deter Iran if its leadership is rational, or at least rational enough not to desire self-destruction. Israeli deterrence would be greatly enhanced, in addition, if the United States were to undertake to provide Israel with a "nuclear umbrella" through guarantees that Washington would treat a nuclear attack on Israel as it would an attack on its NATO allies.

13.5 The Homeland Security Arena: The Impact on Israeli Domestic Preparedness

As noted, it is highly likely that Iran is seeking, at the very least, a potential nuclear capability for reasons having to do with a combination of domestic politics (i.e., regime legitimacy), deterring a theoretical American invasion (both of Iran's immediate neighbors have, of course, been previously occupied by the US military), and enhancing Iranian prestige and "leadership" in the Middle East and beyond—as opposed to a real desire to destroy Israel through nuclear war. Nevertheless, Israeli planners cannot avoid thinking about a doomsday scenario in which Iran

[10] Boeing Defense, Space and Security [7].

[11] Bergman et al. [8]; The Nuclear Threat Initiative [9].

[12] Beres and Chain [10].

would want to attack Israel with nuclear weapons, Israeli (and American) efforts at deterrence were to fail, and Israel's active anti-ballistic missile defenses were to prove incapable of providing a safety net of 100 % and thus that one or two warheads would reach their respective targets. While this is an extremely low probability event, the consequences are clearly so devastating that they cannot be ignored in terms of planning for such a scenario.

Somewhat ironically, while Israel has had to deal with wars on a fairly regular basis throughout its history, investment in civil defense has traditionally been uneven. Early on in Israel's history, apartment buildings, private homes, and public spaces were equipped with bomb shelters and civil defense measures were put in place, but the reality of Israel's air dominance over Arab-Israeli battlefields has meant that the Israeli population never really had to worry about bombing runs by any of the Arab air forces and consequently, as some public air raid shelters became havens for drug addicts while others became storage spaces.[13]

With the development of missile technology however, Israel's Arab adversaries found a way to circumvent Israel's air dominance and thus to strike directly at the Israeli population in their cities. This was brought home to Israel in 1991 when Saddam Hussein fired 39 modified SCUD B missiles at the city of Haifa and the metropolitan Tel Aviv area. The threat of these missiles was made more impactful for the Israeli populace by the fact that it was feared that the missiles would be carrying chemical warheads and consequently the population was provided gas masks and atropine injections (to counter nerve agents) and was trained in their use. Ultimately, the dreaded chemical warheads never materialized and the conventional SCUDs produced only minimal damage—in part due to Coalition Forces missile-hunting activity in western Iraq.[14]

The 1991 Gulf War led, however, to the growing realization in Israel that the enemy was shifting its focus from engaging the Israeli military on the battlefield to taking advantage of rocket and missile technology to hit Israel's cities. This effectively meant that the battle space was shifting to the home front. In 1992, the Israeli military, which had hitherto operated within for regional commands (North, Central, and South) added an additional regional command, the Homefront Command (HFC).

In peacetime, the HFC is responsible for establishing emergency procedures, supervising preparedness exercises and monitoring the preparedness of the first-responder community, health system, local government, the transportation system, utilities, etc., and it creates the doctrine and rules of engagement under which all the agencies responsible for response and recovery operate.[15] The HFC is responsible not only for creating a common response and recovery doctrine and training and exercising, but also for operational national warning systems (sirens, cell phone texting, etc.), instructing the population on emergency measures, command and

[13]Prince-Gibson [11].

[14]Rosenau [12].

[15]Morag [13], p. 213.

control of response agencies during a Special Homefront Situation, the distribution of gas masks and other protective equipment to the population, and the maintenance of bomb shelters.

It is important to bear in mind that the HFC was created in the wake of missile attacks on Israel's population centers and with the threat of chemical warheads in mind. It has since been activated during Israel's war with Hezbollah in 2006 as well as Israel's operations against Hamas in the Gaza Strip in 2006, 2008, 2012, and 2014 due to massive rocket attacks, albeit with conventional warheads.

With the disappearance of Iraq as a military power in the wake of the 2002 US invasion and with the collapse of Syria into civil war and the Assad government's agreement to decommission its chemical weapons arsenal as a result of international pressure in the wake of chemical weapons attacks against civilians and anti-regime rebels in August 2013, there have been reports that Israel is considering cancelling its costly program of providing gas masks/atropine kits to the population.[16] This suggests that at least some members of the defense establishment in Israel do not consider a nuclear attack on the part of Iran something that is worth considering in the foreseeable future, in terms of the budgeting and planning cycle, because gas masks can be useful under certain conditions in protecting people from nuclear fallout. However, it is far from clear yet as to whether gas masks will be phased out in the long term.

According to a 2009 study by the Center for Strategic and International Studies, the effect on the Tel Aviv metropolitan area of a 20 kt explosion (larger than the Hiroshima bomb but slightly smaller than the Nagasaki bomb) would be approximately 50,000 fatalities within a week and about 150,000 within a year. A larger 100 kt warhead would result in an estimated 400,000 fatalities within a week and over 600,000 fatalities within a year.[17] Clearly these are unimaginable and inconceivable losses for any country, particularly one the size of Israel... and this scenario involves just one successful attack.

The scenario of an Iranian nuclear attack against one or more Israeli cities would, however, not necessarily require a completely different approach to civil defense. Some 50 % of the energy released in a nuclear explosion is released through the pressure wave. Depending on the distance from ground zero and the degree to which structures can withstand tremendous wind pressures, people within structures have a chance of surviving the blast (and most construction in Israel involves steel reinforced concrete, which is certainly more survivable than wood). Another 35 % of the energy of the blast is released in the form of heat and, in this case as well, depending on distances and other factors, reinforced structures can provide protection from the kind of burns that would be experienced by people who were unlucky enough to be outside or in wood structures. Finally, only an estimated

[16]"Defense Establishment Weights Cancellation of Further Gas Mask Distribution," *The Jerusalem Post*, December 11, 2013, available at: http://www.jpost.com/Defense/Defense-establishment-weighs-cancellation-of-further-gas-mask-distribution-331512 (accessed September 3, 2013).

[17]Toukan and Cordesman [14], slide 11.

15 % of energy from a nuclear explosion takes the form of ionized radiation, with three-quarters of that radiation lingering for up to several years.[18] In this case, the use of gas masks, protective clothing, and other measures can reduce fatalities from radiation poisoning and, downstream, from cancer, as well as minimize the phenomenon of radiation-induced birth defects, sterility, etc.

For Israeli leaders, moreover, there has reportedly been a tunnel built in the foothills of Jerusalem, and incorporated into civil defense drills since 2006, which is designed to allow the leadership to survive a nuclear attack, respond to that attack, and run the civil defense effort.[19] In short, the Israeli defense establishment does consider a WMD scenario, including one using nuclear weapons, to be conceivable and has taken at least some steps to try to cope with such an unpalatable scenario despite the fact that shorter-range budgetary considerations may lead to a temporary cessation of the provision of gas masks and/or other protective equipment to the population.

13.6 Conclusion

From the Israeli perspective, the ball is primarily in Iran's court as there is little that Israel can do, one way or the other, to control the outcome of events. Teheran may actually not be desirous of building nuclear weapons, as opposed to maintaining the capacity to do so, for the political purposes touched on earlier, and thus Israeli apprehensions with respect to a nuclear attack may be overblown. Even if the Iranians do build one or more nuclear devices and successfully fit these to their Shahab-3 intermediate range missile, the likelihood of Iran actually using such a weapon is extremely low, both because there is little evidence to suggest that the Iranian regime is irrational (given the Israeli, and perhaps American, response to an Iranian nuclear attack on Israel) and because Israel is simply not important enough to Iran and Iranian interests, to risk all in a direct nuclear conflict with it. Of course, the Israeli authorities need to prepare for low risk but high consequence scenarios because a nuclear attack would be absolutely devastating for a small country like Israel, and hence the security establishment in Israel needs to plan accordingly and seek to both prevent successful attacks through active defenses and cope with the aftermath of successful attacks, as horrific as that scenario is to consider. What is more likely however, is that an Iranian capacity to possess nuclear weapons will be more impactful on Israeli regional interests and Israel's struggles with Hamas, Hezbollah, and its other immediate regional rivals, because itwill enhance

[18]Ibid., slide 33.

[19]"Israeli Leaders Spend Day in 'Nation's Tunnel' Nuclear Bunker," *The Telegraph*, June 22, 2011, available at: http://www.telegraph.co.uk/news/worldnews/middleeast/israel/8592107/Israeli-leaders-spend-day-in-Nations-Tunnel-nuclear-bunker.html (accessed September 3, 2014).

the perception of Iranian power and may embolden the Islamic Republic to take greater risks and support even more aggressive actions against Israel, via its proxies and allies, in the future.

References

1. Hayden M, Bayh E (2014) A doubly dangerous Iran. Wall St J :A11. Available at: http://www.wsj.com/articles/michael-hayden-and-evan-bayh-a-doubly-dangerous-iran-1405638861. Accessed 31 Aug 2014
2. Hashem A (2014) How Haider al-Abadi became Iraq's next prime minister. AL-Monitor, 14 August 2014. Available at: http://www.al-monitor.com/pulse/originals/2014/08/iraq-iran-prime-minister-alabadi-maliki.html. Accessed 31 Aug 2014
3. Frantzman SJ (2014) Terra incognita: Nasrallah's dangerous game. Jerusalem Post, 17 August 2014. Available at: http://www.jpost.com/Opinion/Columnists/Terra-incognita-Nasrallahs-dangerous-game-371358. Accessed 31 Aug 2014
4. The White House (2013) Fact sheet: first step understandings regarding the Islamic Republic of Iran's nuclear program. The White House, Office of the Press Secretary, Washington, DC. Available at: http://www.whitehouse.gov/the-press-office/2013/11/23/fact-sheet-first-step-understandings-regarding-islamic-republic-iran-s-n. Accessed 3 Sept 2014
5. Kimball DG (2014) The Iranian uranium-enrichment challenge. Arms Control Today, vol 44, June. Available at: http://www.armscontrol.org/act/2014_06/Focus/The-Iranian-Uranium-Enrichment-Challenge. Accessed 3 Sept 2014
6. Zanotti J, Katzman K, Gertler J, Hildreth SA (2012) Israel: possible military strike against Iran's nuclear facilities, CRS report for congress, R42443. Congressional Research Service, Washington, DC, p 37
7. Boeing Defense, Space and Security (2013) Backgrounder: Arrow weapon system. Boeing Defense, Space and Security, St. Louis. Available at: http://www.boeing.com/assets/pdf/defense-space/space/arrow/docs/Arrow_overview.pdf. Accessed 3 Sept 2013
8. Bergman R, Follath E, Keinan E, Nassauer O, Schmitt J, Stark H, Wiegold T, Wiegrefe K (2012) Operation Samson: Israel's deployment of nuclear missiles on subs from Germany. Der Spiegel Online, 4 June 2012. Available at: http://www.spiegel.de/international/world/israel-deploys-nuclear-weapons-on-german-built-submarines-a-836784.html. Accessed 2 Sept 2014
9. The Nuclear Threat Initiative (2013) Country profiles: Israel. NTI, Washington, DC. Available at: http://www.nti.org/country-profiles/israel/delivery-systems/. Accessed 2 Sept 2014
10. Beres LR, Chain JT (2014) Living with Iran: Israel's strategic imperative, BESA Center Perspectives Paper, No. 249, 28 May 2014
11. Prince-Gibson E (nd) From Israel, lessons in civil defense. Washington Post. Available at: http://www.washingtonpost.com/wp-srv/health/specials/preparedness/israel.html. Accessed 3 Sept 2014
12. Rosenau W (2002) Special operations forces and elusive enemy ground targets, Chp. 3. RAND Corporation, Santa Monica. Available at: http://www.rand.org/content/dam/rand/pubs/monograph_reports/MR1408/MR1408.ch3.pdf. Accessed 11 Aug 2014
13. Morag N (2011) Comparative homeland security: global lessons. Wiley, New York, p 213
14. Toukan A, Cordesman A (2009) Iran, Israel and the effects of a nuclear conflict in the Middle East. Center for Strategic and International Studies, Washington, DC

Chapter 14
Nuclear Security Regulation in the United States: An Overview of U.S. NRC Functions and Activities

Jure Kutlesa

Abstract The United States Nuclear Regulatory Commission (NRC) was created as an independent agency by the United States Congress in 1974 to ensure the safe use of radioactive materials for beneficial civilian purposes. Both the Atomic Energy Act of 1954 and the Energy Reorganization Act of 1974 give the NRC the responsibility for ensuring that the peaceful uses of nuclear energy "make the maximum contribution to the common defense and security and the national welfare." To achieve these ends, the NRC regulates commercial nuclear power plants and other uses of nuclear materials through licensing, inspection and enforcement of its requirements. These requirements are based on U.S. administrative law. In the field of nuclear security, the NRC regulates the following aspects of nuclear power plant activity: Physical Security; Access Authorization; Access Control; Security Equipment; Protective Strategy; Security Training; and Fitness for Duty. This paper will give an overview of the U.S. NRC security related regulatory functions and activities and explain how the NRC achieves its mission of "promoting the common defense and security" specifically as it relates to nuclear reactors.

14.1 Atomic Energy: From War to Peace

July 1945 inaugurated a new era in warfare. The first test of a nuclear weapon (code named "Trinity") was conducted in New Mexico. The development of nuclear weapons was a strictly guarded military secret for the United Sates. After World War II, the knowledge of the destructive potential of nuclear energy was all too apparent.

Soon after, a debate began in the United States between the military and civilian authorities over the control of this technology. Numerous hearings were held in Congress which resulted in Senator Brien McMahon (D-Connecticut) introducing a bill on December 20, 1945 that placed the control of the nation's atomic energy

J. Kutlesa (✉)
Nuclear Regulatory Commission, Washington, DC, USA
e-mail: jure.kutlesa@nrc.gov

S. Apikyan, D. Diamond (eds.), *Nuclear Threats and Security Challenges*,
NATO Science for Peace and Security Series B: Physics and Biophysics,
DOI 10.1007/978-94-017-9894-5_14

program in civilian hands. President Truman signed the McMahon Act, known officially as the Atomic Energy Act of 1946, on August 1, 1946. With this law a new civilian agency, the United States Atomic Energy Commission (AEC), was created.[1]

The Atomic Energy Act of 1946 stated that "the significance of the atomic bomb for military purposes is evident. The effect of the use of atomic energy for civilian purposes upon the social, economic, and political structures of today cannot now be determined".[2]

With this focus on the military uses of nuclear power the Act did not allow for the private use of nuclear material. Instead, the Act's purpose included a program of federally conducted research and development to assure the Government of adequate scientific and technical accomplishment and a program for Government control of the production, ownership, and use of fissionable material to assure the common defense and security.[3]

By 1953 the realities of the Cold War were becoming evident. Both the United States (U.S.) and the Union of Soviet Socialist Republics (USSR) had tested numerous atomic weapons. While meeting with the leaders of the United Kingdom and France in Bermuda on December 4, 1953, U.S. president Dwight D. Eisenhower stated the world was in a rather hysterical condition about the atomic bomb.[4] Eisenhower had been invited by the United Nations to give a speech at their General Assembly and wanted to discuss an idea addressing this concern.

The declassified minutes of that meeting have Eisenhower continuing with his idea that he felt something might be said which, while admitting the terrible destructive quality of atomic energy, might express the point of view of the free world on the constructive capabilities of atomic energy.[5] Four days later, Eisenhower gave an address to the General Assembly of the United Nations in New York City. The subject was the "Peaceful Uses of Atomic Energy". Also known as the "Atoms for Peace" speech, Eisenhower presented a framework for harnessing the peaceful use of nuclear power:

> The United States would seek more than the mere reduction or elimination of atomic materials for military purposes.
>
> It is not enough to take this weapon out of the hands of the soldiers. It must be put into the hands of those who will know how to strip its military casing and adapt it to the arts of peace.

---

[1] Department of Energy, Office of History and Heritage Resources [1].

[2] ATOMIC ENERGY ACT OF 1946,(Public Law 585, 79th Congress) http://www.osti.gov/atomicenergyact.pdf. Accessed 19 Aug 2014.

[3] Ibid. Accessed 19 Aug 2014.

[4] DDE's Papers as President, International Meetings Series, Box 1, Bermuda-State Dept. Report-Top Secret. MemorandumofConversationregardingBermudaMeeting,December4,1953. http://www.eisenhower.archives.gov/research/online_documents/atoms_for_peace/Binder4.pdf Accessed 19 Aug 2014.

[5] Ibid. Accessed 19 Aug 2014.

> The United States knows that if the fearful trend of atomic military buildup can be reversed, this greatest of destructive forces can be developed into a great boon, for the benefit of all mankind.
>
> The United States knows that peaceful power from atomic energy is no dream of the future. That capability, already proved, is here – now – today. Who can doubt, if the entire body of the world's scientists and engineers had adequate amounts of fissionable material with which to test and develop their ideas, that this capability would rapidly be transformed into universal, efficient, and economic usage.[6]

Although appearing altruistic, the facts behind sharing nuclear knowledge have to be viewed through the strategic realities of the Cold War. The prevailing sense of urgency, at least among U.S. Government leaders, reflected the fear of falling behind other nations in fostering peaceful atomic progress. The strides that Great Britain was making in the field seemed disturbing enough, but the possibility that the USSR might surpass the United States in civilian power development was even more ominous. AEC Commissioner Thomas E. Murray described a "nuclear power race" in a 1953 speech and warned that the "stakes are high".[7]

In order to shift the emphasis of atomic energy from military to peaceful uses, United States law needed to be changed. The Atomic Energy Act of 1946 with its focus on military use and government control of nuclear information and materials was amended. The resulting legislation, the Atomic Energy Act of 1954 (the Act), was signed by President Eisenhower on August 30, 1954. In his statement upon signing the new law, Eisenhower said, "As I sign this bill, I am confident that it will advance both public and private development of atomic energy – that it will thus lead to greater national strength – and that programs undertaken as a result of this new law will help us progress".[8]

The goal of the Act was:

> Atomic energy is capable of application for peaceful as well as military purposes. It is therefore declared to be the policy of the United States that the development, use, and control of atomic energy shall be directed so as to make the maximum contribution to the general welfare, subject at all times to the paramount objective of making the maximum contribution to the common defense and security; and the development, use, and control of atomic energy shall be directed so as to promote world peace, improve the general welfare, increase the standard of living, and strengthen free competition in private enterprise.[9]

The Act instructed the AEC to prepare regulations that would protect public health and safety from radiation hazards. Thus, it assigned the agency three major functions: (1) to continue its weapons program, (2) to promote the commercial uses of nuclear power, and (3) to protect against the hazards of those peaceful applications.[10]

---

[6] American Presidency Project [2].

[7] Walker and Wellock [3].

[8] Eisenhower [4].

[9] THE ATOMIC ENERGY ACT OF 1954, AS AMENDED (Public Law 83–703). http://pbadupws.nrc.gov/docs/ML1327/ML13274A489.pdf#page=23. Accessed 20 Aug 2014.

[10] Walker and Wellock [5].

14.2 New Agencies: NRC & DoE

As time went on the three competing AEC functions became increasingly difficult to manage. The AEC came under attack for the conflicting responsibilities of promoting the development of civilian nuclear power and regulating the same. After the energy crisis of the early 1970s, then President Richard Nixon asked the U.S. Congress to create a new agency whose only responsibility was to regulate and license the commercial use of nuclear material.

In 1974 the U.S. Congress passed the Energy Reorganization Act which eliminated the Atomic Energy Commission and created the Nuclear Regulatory Commission (NRC) and the Energy Research and Development Agency (ERDA). The NRC was created to regulate the commercial nuclear industry while ERDA managed nuclear research, development, nuclear weapons and naval reactors. Two years later, in 1977, ERDA was merged with several other agencies to create the U.S. Department of Energy.[11]

14.3 Legal Basis for NRC Activities

So far we have discussed the history of the Nuclear Regulatory Commission. Now we turn to the legal basis for the civilian control of nuclear material in the United States.

The U.S. Constitution organized the federal government into three branches: Article 1: Legislative (Congress); Article 2: Executive (President); Article 3: Judiciary (Courts). Each branch of government can make laws. The legislative branch creates statutory law; the executive branch creates regulatory law; and the judicial branch creates common law.

As we have seen, the fundamental law establishing the creation and purpose of the NRC is the Atomic Energy Act of 1954. Since the U.S. Congress created this law it is considered a statute. However, through the years since its first implementation, the Act has been amended numerous times. Therefore in legal documents we refer to the Act as "The Atomic Energy Act of 1954, as amended".

The second important law that affects the NRC is the Energy Reorganization Act of 1974 which created the NRC and specified its functions in relation to control of civilian use of nuclear material.[12] The Reorganization Act sates that:

> There is established an independent regulatory commission to be known as the Nuclear Regulatory Commission which shall be composed of five members, each of whom shall be a citizen of the United States... Members of the Commission shall be appointed by the President, by and with the advice and consent of the Senate... Appointments of members

[11]Department of Energy, Office of Management [6].

[12]Energy Reorganization Act of 1974, (Public Law 93-438). http://www.gpo.gov/fdsys/pkg/STATUTE-88/pdf/STATUTE-88-Pg1233.pdf. Accessed 25 Sep 2014.

> pursuant to this subsection shall be made in such a manner that not more than three members of the Commission shall be members of the same political party. Each member shall serve for a term of five years . . . [13]

As a result, the NRC is an independent agency by virtue of the Reorganization Act. It does not belong to any other department in the federal government. The NRC is part of the executive branch insofar as the president appoints the commissioners. However, the Reorganization Act states that any commissioner may be removed by the President only for inefficiency, neglect of duty, or malfeasance in office.[14] This establishes both the independence of the NRC and its executive agency status.

In the Reorganization Act of 1974, Congress directs the NRC to regulate certain activities in order to ensure radiological health and safety, and common defense and security.[15] As a result, Congress gives the NRC authority to issue rules and regulations to help carry out NRC's mission. Rules are administrative laws created by the Executive branch of the federal government in order to implement statutes created by Congress. These laws encompass both legally binding requirements (regulations and orders) and non-binding guidance.

In an effort to provide guidance as to how an agency may create administrative law, the U.S. Congress adopted the Administrative Procedure Act of 1946 (APA). This statute governs the way agencies propose and establish regulations.[16] The APA requires agencies to inform the People of these new rules and directs their publication the Federal Register (the official journal of the U.S. Federal government). In an effort to organize rules into a an easily accessible format, once published in the Federal Register these rules are then compiled in the Code of Federal Regulations (CFR) (there are 50 titles that cover broad subject matters and the NRC rules are found in Title 10 – Energy).[17]

Located in Title 10, Code of Federal Regulations, Sections 73.55 and 73.56 (abbreviated as 10 CFR 73.55 & 73.56) are the specific rules that govern security at commercial nuclear power plants.

14.4 NRC Rules for Power Reactor Security

10 CFR 73.55, entitled "Requirements for Physical Protection of Licensed Activities in Nuclear Power Reactors Against Radiological Sabotage", states that power reactors "shall establish and maintain a physical protection program, to include a

[13] Ibid. Accessed 25 Sep 2014.

[14] Ibid. Accessed 25 Sep 2014.

[15] Ibid. Accessed 25 Sep 2014.

[16] Administrative Procedure Act of 1946, (Public Law 79-404). http://www.legisworks.org/congress/79/publaw-404.pdf. Accessed 26 Sep 2014.

[17] Code of Federal Regulations: Government Printing Office. http://www.gpo.gov/fdsys/browse/collectionCfr.action?collectionCode=CFR. Accessed 26 Sep 2014.

security organization, which will have as its objective to provide high assurance that activities involving special nuclear material are not inimical to the common defense and security and do not constitute an unreasonable risk to the public health and safety".[18]

Additionally, 10 CFR 73.1 states that:

The following design basis threats ... shall be used to design safeguards systems to protect against acts of radiological sabotage and to prevent the theft or diversion of special nuclear material.

1. Radiological sabotage

   (i) A determined violent external assault, attack by stealth, or deceptive actions, including diversionary actions, by an adversary force capable of operating in each of the following modes: A single group attacking through one entry point, multiple groups attacking through multiple entry points, a combination of one or more groups and one or more individuals attacking through multiple entry points, or individuals attacking through separate entry points, with the following attributes, assistance and equipment:

      (a) Well-trained (including military training and skills) and dedicated individuals, willing to kill or be killed, with sufficient knowledge to identify specific equipment or locations necessary for a successful attack;
      (b) Active (e.g., facilitate entrance and exit, disable alarms and communications, participate in violent attack) or passive (e.g., provide information), or both, knowledgeable inside assistance;
      (c) Suitable weapons, including handheld automatic weapons, equipped with silencers and having effective long range accuracy;
      (d) Hand-carried equipment, including incapacitating agents and explosives for use as tools of entry or for otherwise destroying reactor, facility, transporter, or container integrity or features of the safeguards system; and
      (e) Land and water vehicles, which could be used for transporting personnel and their hand-carried equipment to the proximity of vital areas; and

   (ii) An internal threat; and
   (iii) A land vehicle bomb assault, which may be coordinated with an external assault; and
   (iv) A waterborne vehicle bomb assault, which may be coordinated with an external assault; and
   (v) A cyber attack.

It is within these parameters, and others further specified in non-public documents, that nuclear power plants in the U.S. must defend.

[18] 10 CFR 73.55. http://www.nrc.gov/reading-rm/doc-collections/cfr/part073/part073-0055.html. Accessed 26 Sep 2014.

14.5 The U.S. Approach to Defense in Depth

The NRC has developed its regulatory requirements by beginning with the idea of defense in depth. Defense in depth refers to multiple layers of security that can both stand alone and complement each other. For example, nuclear power plants are physically designed to be robust structures that can withstand hurricanes, tornadoes, and earthquakes. This robust design additionally helps in securing the plant from adversaries attempting to attack a plant.

The NRC requires that each power plant have a security force that is well-trained and armed and physical barriers that prevent the intrusion of both personnel and vehicles. To gain entry to power plants, one needs to have completed a thorough background check and then, if authorized, must enter through an access control area that searches and screens all individuals as well as their possessions. Plants must have advanced intrusion detection systems that are continuously monitored from at least two alarm stations, both of which must be robust structures.

The plant must also protect against an insider threat by employing a Behavior Observation Program which allows for the detection of aberrant behavior in employees and visitors. This program is designed to detect behaviors or activities that may constitute an unreasonable risk to the health and safety of the public and common defense and security, including a potential threat to commit radiological sabotage.

Additionally, an insider mitigation program requires security personnel to monitor the initial and continuing trustworthiness and reliability of individuals granted or retaining access to the plant and to "implement defense-in-depth methodologies to minimize the potential for an insider to adversely affect, either directly or indirectly, the plant's capability to prevent significant core damage and spent fuel sabotage".[19]

An additional layer of protection is in place which coordinates threat information and response. The NRC works closely with Federal, State, Tribal, and local authorities. These relationships help ensure that the NRC can act quickly to disseminate threat information to licensees and allow effective emergency response in the event of an attack. The NRC also actively works with the U.S. national intelligence community. NRC personnel review and analyze intelligence information and communicate with other intelligence organizations. Together, these layers of defense provide the highest level of security in the U.S. commercial sector.

14.6 Regulation: In Practice

In practice, the NRC regulations fall under two distinct categories: the "performance based" approach and the "compliance based" approach.

[19] 10 CFR 73.56 http://www.nrc.gov/reading-rm/doc-collections/cfr/part073/part073-0056.html. Accessed 29 Sep 2014.

The performance based approach focuses on desired, measurable outcomes, rather than prescriptive processes, techniques, or procedures. Performance-based regulation leads to defined results without specific direction regarding how those results are to be obtained. At the NRC, performance-based regulations focus on identifying performance measures that ensure an adequate security margin and offer incentives for nuclear power plants to improve security without formal regulatory intervention by the agency. In other words, the NRC states a goal to achieve but does not direct how to implement it.

An example of a performance based regulation is the Force on Force (FoF) inspection. An FoF inspection is a two-phased, performance-based inspection that is designed to verify and assess the ability of the nuclear power plant to defend itself against a mock adversary force. An FoF inspection is conducted over several days in two separate visits by an inspection team at the power plant. Along with the NRC inspectors, active-duty U.S. Special Operations Forces advise the NRC inspection teams that conduct force-on-force inspections. These individuals participate in the inspections by helping the NRC inspectors develop the scenarios, providing expert technical advice to the mock adversary force, and assisting the NRC inspectors in evaluating site security forces and systems.[20]

On the other hand, "Compliance-based" regulations direct the power plant to strictly meet security criteria. Certain fundamental requirements (a planned response by the power plant security force, specified heights of fencing, etc.) ensure that the foundation of a reliable and redundant security system is in place.

An example of a compliance based regulation is the requirement that a "fence be constructed of No. 11 American wire gauge, or heavier wire fabric, topped by three strands or more of barbed wire or similar material on brackets angled inward or outward between 30 and 45 from the vertical, with an overall height of not less than eight feet, including the barbed topping".[21] Compliance-based regulation has limitations since it is not possible to identify and enforce all the necessary security requirements on the regulatory side. Therefore the compliance based approach is complimented by the performance based regulatory approach.

14.7 Inspectors

In the United States there are currently 100 nuclear power plants operating at 62 commercial reactor sites.[22] Two employees from the NRC work at each of these sites. These personnel are called "resident inspectors". Their job is to monitor the power plants safe and secure operation and compliance with regulatory require-

[20]USNRC [7].

[21]10 CFR 73.2 http://www.nrc.gov/reading-rm/doc-collections/cfr/part073/part073-0002.html. Accessed 2 Oct 2014.

[22]USNRC [8].

ments. Resident inspectors are nuclear engineers who have additional training in other applicable disciplines such as health physics and security. Resident inspectors are supported by technical specialists located at one of the four NRC Regional Offices. These specialists include nuclear security inspectors.

Nuclear security inspectors are technical security specialists with years of specialized training in both security and nuclear power plant operations. Nuclear security inspectors conduct baseline security inspections at the power plants throughout the year. The inspections cover the following topics: Physical Security; Access Authorization; Access Control; Security Equipment; Protective Strategy; Security Training; Force on Force; and Fitness for Duty.

Nuclear security inspectors are not limited to inspecting only the above topics. They use their broad knowledge to ensure the power plants are complying with regulatory requirements and fundamental nuclear security. Additionally, nuclear security inspectors support the resident inspectors when they have technical questions beyond the scope of their knowledge.

14.8 Conclusion

The United States Nuclear Regulatory Commission's mission is to regulate the nation's civilian use of byproduct, source, and special nuclear materials to ensure adequate protection of public health and safety, to promote the common defense and security, and to protect the environment.[23]

The NRC traces its history to the first days of the nuclear age with the U.S. government's creation of the Atomic Energy Commission. Under the Atomic Energy Act of 1954, a single agency, the Atomic Energy Commission, had responsibility for the development and production of nuclear weapons and for both the development and the safety regulation of the civilian uses of nuclear materials. The Energy Reorganization Act of 1974 split these functions, assigning to one agency, now the Department of Energy, the responsibility for the development and production of nuclear weapons, promotion of nuclear power, and other energy-related work, and assigning to the NRC the regulatory work, which does not include regulation of defense nuclear facilities.

The NRC's regulatory mission covers three main areas: reactors, materials and waste. This paper focused on the security aspects of commercial reactors used for generating electric power at nuclear power plants.

The NRC demands high assurance from power plant operators that they will protect their sites against adversaries attempting to commit radiological sabotage. Through its regulatory regime and long history, the U.S. NRC has ensured the safe civilian use of nuclear materials.

[23]USNRC Mission Statement. http://www.nrc.gov/about-nrc.html. Accessed 30 Sep 2014

Disclaimer Mr. Kutlesa contributed this article in his personal capacity. The views expressed are his own and do not necessarily represent the views of the Nuclear Regulatory Commission or the United States Government.

References

1. Department of Energy, Office of History and Heritage Resources (2014) https://www.osti.gov/manhattan-project-history/Events/1945-present/civilian_control.htm. Accessed 19 Aug 2014
2. American Presidency Project. Dwight D. Eisenhower. In: 256 – Address before the General Assembly of the United Nations on Peaceful Uses of Atomic Energy. New York City. http://www.presidency.ucsb.edu/ws/?pid=9774. Accessed 20 Aug 2014
3. Walker J, Wellock T (2010) A short history of nuclear regulation: 1946–2009. US NRC, Rockville
4. American Presidency Project. Dwight D. Eisenhower. In: 223 – Statement by the President upon signing the Atomic Energy Act of 1954
5. Walker JS, Wellock TR (2010) A short history of nuclear regulation, 1946–2009, NUREG/BR-0175, revision 2. US NRC, Rockville, pp 3–4
6. Department of Energy, Office of Management (2014) http://energy.gov/management/office-management/operational-management/history/doe-history-timeline. Accessed 20 Aug 2014
7. USNRC (2011) Protecting our nation: a report of the U.S. Nuclear Regulatory Commission (NUREG/BR-0314, Rev. 2). http://www.nrc.gov/reading-rm/doc-collections/nuregs/brochures/br0314/. Accessed 30 Sep 2014
8. USNRC (2014) 2014–2015 information digest (NUREG-1350, vol 26). http://www.nrc.gov/reading-rm/doc-collections/nuregs/staff/sr1350/v26/sr1350v26.pdf. Accessed 30 Sep 2014

Chapter 15
Current Activities of the European Union in Fighting CBRN Terrorism Worldwide

Jozef Sabol, Bedřich Šesták, Lubomír Polívka, and Kamil Mroz

Abstract The paper describes the development and present situation related to establishment of the regional Chemical, Biological, Radiological and Nuclear (CBRN) Centers of Excellence (CoEs) under the European Union (EU) initiative Instrument for Stability (IfS) in various parts of the world outside Europe in order to assist the EU in fighting the international CBRN threats and to minimize the associated hazards by adopting adequate prevention, preparedness and response measures. The main aim of CoEs and relevant Regional Secretariats is to strengthen the institutional capacities of selected countries to mitigate CBRN risks, including such criminal activities as CBRN proliferation and terrorism. The CoEs are supposed to assist the EU in developing a structural, all-hazards CBRN policy at the national, regional and international levels to be able to respond the these threats, and to reduce the vulnerability of countries to CBRN events. Until recently, main attention in this field was paid to the countries of the former Soviet Union focusing on nuclear safeguard and security of nuclear material and high-activity radioactive sources. At present, however, the growing interest in developing nuclear, biotechnology and chemical capabilities in many countries in Africa, Middle East and in South East Asia requires the extension in implementing reliable and efficient tools to improve the safety and security of the CBRN material on a global scale. The presentation also includes some personal observations and impressions of one of the authors (JS) who, under the EU program, visited the CoEs in Manila (Philippines), Tashkent (Uzbekistan), Amman (Jordan) and Rabat (Morocco).

In addition, some aspects related to the preparedness of the Czech Republic for a nuclear or radiological emergency including a radiological terrorist attack have also been discussed.

J. Sabol (✉) • B. Šesták • L. Polívka
Police Academy, Prague, Czech Republic
e-mail: sabol@polac.cz

K. Mroz
JCI, Wallonia, Belgium

S. Apikyan, D. Diamond (eds.), *Nuclear Threats and Security Challenges*, NATO Science for Peace and Security Series B: Physics and Biophysics, DOI 10.1007/978-94-017-9894-5_15

15.1 Introduction

Some of the CBRN materials or agents are used in a number of civilian applications where there is not in all cases sufficient control over these dangerous substances which may principally be used for the construction of Weapons of Mass Destruction (WMD). This is why it is important that all countries introduce adequate measures to control all relevant CBRN materials in accordance with the international requirements and best practices adopted already in most EU member states and in other developed countries. In addition, appropriate measures should be implemented in order to prevent illicit trafficking involving these materials which can be smuggled from countries, with insufficient control of such materials, over borders to other countries and regions where they may subsequently be used for terrorist or other malevolent actions.

In addition to direct assaults on the opponents, the sensitive areas and targets may include well selected places and objects as airports, army installations, subway transport systems and other tunnels, customs and immigration offices, elements of critical national infrastructure, ports, schools and universities, large hospitals, cinemas, theatres, stadiums etc.

Well-known reasons behind the CBRN terrorism are associated with religious, ethnic, racial, political and territorial disputes and intolerance. In these cases, it is typically members of orthodox groups who use CBRN terrorism to try to impose their particular interests and goals using violent and malicious means while themselves being prepared to die and as take innocent victims.

15.2 CBRN Materials Present a Global Threat

The events of 9/11 (September 11th 2001-attack on the World Trade Center, Pentagon and other targets in the USA) [1] have demonstrated the importance of the countries to be prepared for such potential attacks that disregard traditional moral standards, and could kill thousands of civilians without warning. In this coordinated action, the terrorists hijacked four airplanes and deliberately flew them into high-level targets in the USA where about 3,000 people died. These events prompted *the initiation of the "Global War on Terrorism"* (*GWOT*), *carried out by the USA, NATO and other allies against the global threats of terrorism where CBRN materials should be considered as one of real danger in the future potential assaults* (Fig. 15.1).

There have already been a number of attempts to acquire CBRN material, which had been planned for use by terrorist groups. Since the CBRN threats show a global character, the measures needed to be taken against this danger have to be coordinated worldwide and the approach aimed at the prevention, detection and mitigation of consequences of CBRN attacks should be adopted based on a wider international cooperation of all relevant countries and stakeholders.

Fig. 15.1 The attacks which have changed our perception of terrorist threats

In addition to intentional use of CBRN materials (attacks or sabotage), we have to be prepared also for incidents and accidents which may happen in both civilians and military facilities. Successful proliferation prevention, counterforce and active defense operations can reduce the threat, decrease the number of attacks, thwart an attack, and reduce the burden on passive defense and consequence management measures.

The threat of CBRN terrorism is evolving and, with it, the risk of incidents intended to maximize the number of victims on a global scale. There is no doubt that various terrorist groups are working hard to acquire CBRN materials and the expertise to use them in their operations.

15.3 Principle Approach to Minimize the CBRN Danger

To summarize the general differences between biological, radiological, and chemical agents, the effects of chemical and radiological agents are typically recognized within minutes to hours after a release, while it may be anywhere from a couple days to a week before the effects of a biological attack are seen as symptoms (Fig. 15.2). After the attack is recognized as biological, it may take several days to confirm the type of biological agent.

It is relatively difficult to gain access to bio-agents of the sufficient purity and quantity which may have an impact on many people. For radiological agents, large quantities, such as high-activity radioactive sources, are relatively difficult to obtain

AGENTS	SUMMARY CHARACTERISTICS		
	Time To Effects	Potential Impact	Availability
BIO	Days to Weeks	Local to Global	Low
RAD	Minutes to Hours	City to Region	Medium
CHEM	Seconds to Hours	City Blocks	High

Fig. 15.2 Some specific differences between biological (BIO), radiological (RAD) and chemical (CHEM) agents in terms of their effects after the attack (based on the presentation [2])

because of the level of protection and security recently introduced in most countries. Smaller quantities, such as those found in commercial or university equipment, may be easier to access, but these small amounts could not affect a large area. In contrast, many chemical agents are commercially available, making a chemical attack perhaps the easiest method to carry out.

Common health effects associated with released chemical agents include dizziness, nausea, blindness, and disorientation. High enough concentrations or amount of these agents can cause serious injury, immobilization, and even death. The most immediate and simple mitigation measure is to move away from the chemical agent release source or cloud, filter breathing, even if this is just using a crude system such as a handkerchief, and reach fresh air. Then remove clothes, clean skin and eyes, and administer an antidote if available and applicable.

Bio-Agents (bacteria, viruses, and bio-toxins) used for malicious purpose may result in wide-spread sickness, death, panic, and to inflict adverse social impact, terror and economic consequences. These agents can cause disease by inhalation, ingestion, or skin contact. Unlike other agents, however, biological agents can also be contagious among people, and between people and animals. Bio-agents may also incubate and multiply in the body for days to weeks before visible symptoms appear and persons recognize that they are sick. There are some bio-agents, for which no vaccines are available, so infected people must be quarantined. Some biological agents, when left in the environment, can be dormant but potent for quite a long period (weeks and even years). This depends on both the specific agent, and on the environmental conditions.

Naturally occurring radiation exists everywhere but is at too low of a concentration to pose a significant threat. However, some artificial radioactive sources of higher activity are suitable for use as a crude radiological weapon, such as a dirty bomb. Such powerful sources can be found in medical radiotherapy facilities, industrial irradiators and some instruments using radioactive sources in such applications as industrial radiography, industrial gauges, soil moisture meters etc.

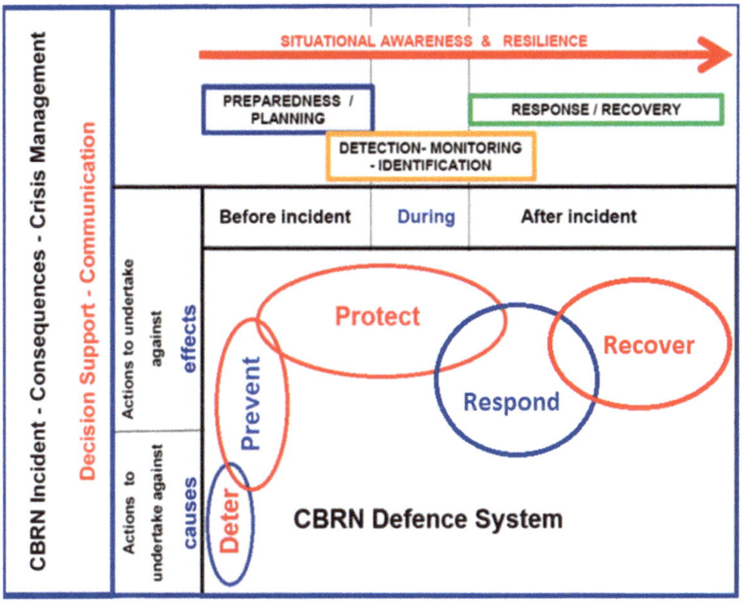

Fig. 15.3 Individual components and phases of the CBRN defence system

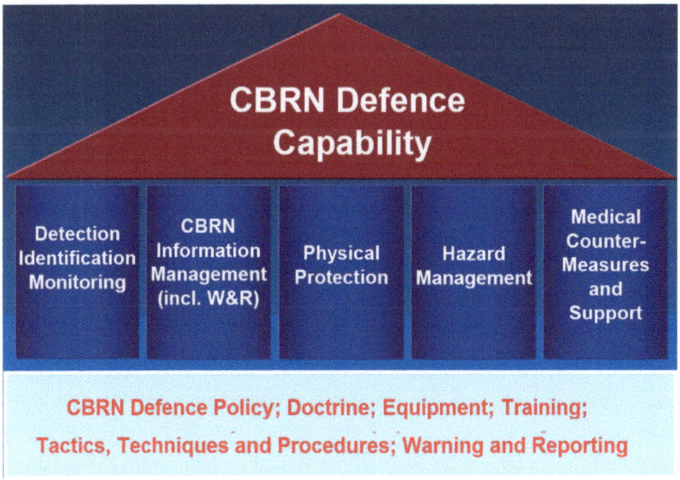

Fig. 15.4 Illustration of main pillars of the CBRN defence capabilities

In order to minimize the potential CBRN danger, several steps are important including deterrence, prevention, protection, response and recovery (Fig. 15.3) [3]. These actions comprise activities before the incident (attack), during the attack and after the attack. The situation can also be illustrated in terms of CBRN defence capabilities which are characterized by seven main pillars (Fig. 15.4) [4].

15.3.1 EU Action Plan to Fight the CBRN

The EU CBRN Action Plan [5] represents the main EU policy document aimed at the protection of the EU Member States against CBRN threats. The Action Plan constitutes a political commitment and may be considered as a roadmap of intentions and measures to be taken in the coming years. It recommends actions concerning prevention, detection, preparedness and response, as well as horizontal measures in the context of high-risk CBRN materials. As to its implementation, the Action Plan should be conducted with full respect for international law, including human rights and the principle of the rule of law.

In the field of biological security, the EU has achieved visible results in providing training and education for professionals working with, having access to or handling high-risk agents, including bio-safety, bio-security and bio-ethics. With respect to radiological protection and nuclear safety/security, the EU Member States reported significant progress in carrying out research into detection and response, including development of technology in the area of protection and response to radiological threats. The work has also been carried out in conducting exercises at local, regional, national, EU and international level, based on risk assessment, establishing the European network of specialised CBRN law enforcement units, setting up the Early Warning System (EWS) for law enforcement authorities for incidents related to high risk CBRN materials as well as explosives and firearms under the coordination of Europol and several Member States.

The EU Action Plan includes some specific preventing measures as (a) Developing of lists at EU level of high-risk CBRN materials; (b) Identifying and reporting suspicious transactions and behavior; (c) Enhancing security and control of high risk CBRN materials; facilities and transport infrastructure; (d) Contributing to the development of a high security culture of staff; improving information exchange, (e) Strengthening the import/export regime; (f) Enhancing the cooperation on the security of CBRN as well as dual materials.

The Plan is also focused on efficient preparedness measures including: (a) Improved emergency planning, (b) Stronger countermeasure capabilities, (c) Upgraded domestic and international information flows regarding CBRN emergencies, (d) The development of new modelling and better decontamination and remediation capacity, and (e) Ensuring greater capacity to conduct criminal investigation.

In accordance with the EU requirements, the responding measures should require: (a) Enhanced international cooperation, (b) Improved lines of communication with the public; (c) More robust information tools for CBRN security, (d) Advanced training courses for first responders, (e) Improved personnel security, and (f) Ensuring that legislation is put in place to tackle CBRN terrorism.

The goals of EU CBRN policy to minimise the threat and damage have also to address the following aspects: (a) Use of a risk-based approach to security, (b) Effective protection of CBRN materials including their storage and transport, (c) Improved exchanges of security-related information among countries, (d) Further development of detection systems, and (e) Provision of the necessary tools to manage CBRN incidents and accidents.

15.3.2 Establishment of EU Centers of Excellence

The EU CBRN Centres of Excellence (CoEs) Initiative [6] was launched in 2010 in response to the need to strengthen the institutional capacity of countries outside Europe to mitigate CBRN risks, including criminal activities (e.g. CBRN proliferation or terrorism), natural disasters (e.g. swine flu) and accidental disasters (e.g. Bhopal or Fukushima). The objective of the CoE Initiative is to develop a structural, all-hazards CBRN policy at the national, regional and international levels to anticipate and respond to these risks, and to reduce the vulnerability of countries to CBRN events. In this respect, the initiative is in the reciprocal interests of regional and EU security.

The initiative addresses the mitigation of and preparedness against risks related to CBRN material and agents. The origin of these risks can be criminal, accidental or natural. The Initiative seeks to boost cooperation at regional and international levels, and to develop a common and coherent CBRN risk mitigation policy at the regional level. Risk mitigation comprises prevention, preparedness and post-crisis management.

The initiative is implemented and funded by the European Commission in cooperation with the Joint Research Centre (JRC), the United Nations Interregional Crime and Justice Research Institute (UNICRI) and a governance team. The Initiative is implemented with the technical support of relevant experts from EU Member States, international organisations and individual experts in the field [7, 8].

The establishment of regional CoEs is a cornerstone of these activities: they offer a coherent and comprehensive approach covering legal, regulatory, enforcement and technical issues. The European Commission (EC) adopted a policy package on CBRN security with the aim to strengthen the protection of EU citizens and to improve the protection against CBRN danger in other regions.

The EU CBRN CoEs are primarily aimed at:

- Networking, regional and international partnerships, consolidating, coordinating and optimising existing capabilities in terms of expertise, training, technical assistance or equipment;
- Addressing regional CBRN needs through specific tailored projects in the fields of concern such as: protection of CBRN material/facilities, public and infrastructure protection, denying support for CBRN misuse and terrorism, border control/border monitoring, export control, transit and trans-shipment control, safeguarding CBRN information diffusion, bio-safety/bio-security, illicit trafficking, CBRN waste management, first response, public health impact mitigation, post incident recovery, investigation and prosecution, crisis response;
- Strengthening a regional culture of safety and security by increasing local ownership, local expertise and long-term sustainability;
- Institutional capacity building at regional and national levels;
- Reinforcing of national CBRN policy, improvement of institutional capacities in legal, regulatory, control, scientific/technical support and law enforcement domains;

- A coherent interagency approach to enhance coordination and effective response;
- Cooperation with international organisations and EU member states to ensure synergy and avoid duplication of efforts;
- Coherence and visibility of the EU action;
- Within the CBRN CoE structure there is a series of Regional Secretariats. The CBRN CoE Regional Secretariats will operate in the following eight regions:
  - African Atlantic Façade;
  - Central Asia;
  - Eastern and Central Africa;
  - Gulf Cooperation Council Countries;
  - Middle East;
  - North Africa;
  - South East Asia; and
  - South East Europe, Southern Caucasus, Moldova and Ukraine.

15.3.3 Coordination with Other Partners Including NATO

NATO experience within CBRN domain is considerable, and there is interest of the EU to enhance cooperation and coordination with NATO, the United States and other countries including China, the Russian Federation and Japan.

The most important issue is to develop and maintain national strategic long term vision which would be based on in depth analysis of global security environment of the twenty-first century. There should be a clear idea of ways and means to counter CBRN threats in the most effective and efficient way.

NATO has been continuing to ensure that the Alliance – its populations, territory and forces – are secure from CBRN threats, including WMD, which require an active political agenda, strong deterrence and defence capabilities, and appropriate civil emergency capabilities. NATO support for arms control, disarmament and non-proliferation will continue to play a major role in its security objectives. NATO is ready to offer necessary support and capabilities to international non-proliferation and CBRN defence efforts, by further developing and harmonising existing military capabilities relevant to preventing WMD proliferation and employing them.

NATO will continue to transform CBRN defence capabilities to reflect the requirement for rapid deployment of mobile, flexible, highly sophisticated forces tailored to the mission and capable of conducting joint and combined operations; the need for increased military and non-military cooperation to address threats; and the infusion of new technology to enhance capabilities.

In implementing this policy, NATO will improve collaboration, as appropriate, with partners, as well as relevant international and regional organisations, and national authorities in member States.

Cooperation of NATO with EU may primarily concentrate on the following agenda:

- Training cooperation with EU;
- Cooperation with EU/EDA;
- Support the EU CBRN Counter-Measures Concept development;
- Cooperate with EU on NATO EU TRG synchronization;
- CBRN Courses organization IAW EUMS request;
- Cooperation with sponsoring nations CoEs; and
- Cooperation with NATO nations.

15.3.4 Situation in the Czech Republic

In the Czech Republic (CR) the framework, which generally governs disaster response and preparedness, applies also to CBRN events [9]. At the central level the National Security Council, which comprises the Prime Minister and other members of the Government and is chaired by the Minister of the Interior, bears the responsibility to coordinate the CR's security issues and prepare draft measures to ensure the country's safety. Crisis management and response concern all bodies and components of the state administration, including regional and local authorities. The operational framework is defined by the civil protection plans, which are controlled by the General Directorate (GD), the Fire Rescue Service (FRS) and the Regional FRS. Operational organisations and civil protection authorities are all part of the GD Fire Rescue Service and perform their tasks as one organisation. The Fire Rescue Service has a unified approach to all types of emergencies and other authorities could be involved according to the kind of crisis at stake.

Preparedness and response to an emergency, including those involving CBRN materials, are regulated by the following laws: The Law on Crisis Management No. 240/2000 and the Law on Integrated Rescue System No. 239/2000. These two laws define the responsibilities of the Government, the central administrative offices and the territorial administrative offices plus elements of the Integrated Rescue System. In addition, they lay down crisis preparedness measures and the derogations to individual rights during a crisis.

Crisis management is considered not only as a set of tasks to crisis preparedness and their solutions, but also as a complex of activities to rehabilitate the infrastructures of the affected territory.

As an EU Member State, the CR Republic is strictly following the EU Action Plan requirements and at the same time has been, for several years, actively involved in the against terrorism in the context of the activities of NATO and the Euro-Atlantic Partnership Council and the Partnership for Peace (EAPC/PfP). The CR and NATO regard CBRN terrorism as one of the principal security threats of the present day. The country has been the leading player in a multinational NATO CBRN battalion. In this context, a CBRN Defence CoE was open in Vyškov in 2004 which provides specialist training for the Armed Forces of the Czech Republic, NATO member countries, partner countries and other countries [10].

The CR has also established a CoE in the Nuclear Research Institute (NRI) Řež, which is specialized in nuclear safety research and nuclear fuel. The first impulse to create the CoE at the NRI was done in the IAEA International Conference on Research Reactors.

15.4 Conclusion

The paper outlined the development and present situation related to establishment of the regional CBRN CoE under the EU initiative Instrument for Stability (IfS) in various parts of the world outside Europe in order to assist the EU in fighting the international CBRN threats and to minimize the associated hazards by adopting adequate prevention, preparedness and response measures. The main aim of CoEs and their Regional Secretariats is to strengthen the institutional capacities of selected countries to mitigate CBRN risks, including such criminal activities as CBRN proliferation and terrorism.

The CoEs are supposed to assist the EU in developing a structural, all-hazards CBRN policy at the national, regional and international levels to be able to respond the these threats, and to reduce the vulnerability of countries to CBRN events. Until recently, main attention in this field was paid to the countries of the former Soviet Union focusing on nuclear safeguard and security of nuclear material and high-activity radioactive sources. At present, however, the growing interest in developing nuclear, biotechnology and chemical capabilities in many countries in Africa, Middle East and in South East Asia requires the extension in implementing reliable and efficient tools to improve the safety and security of the CBRN material on a global scale. The presentation also included some personal observations and impressions of one of the authors (JS) who, under the EU program, visited the CoEs in Manila (Philippines), Tashkent (Uzbekistan), Amman (Jordan) and Rabat (Morocco).

References

1. (2004) The 9/11 Commission report: final report of the National Commission on Terrorist Attacks upon the United States including the executive summary, National Commission on Terrorist Attacks, W.W. Norton and Company, Inc., New York. ISBN 0-393-06041-1
2. McLeon VA An introduction to chemical, biological and radiological threat agents. Online: https://www.google.cz/#q=An+Introduction+to+Chemical%2C+Biological%2C+ppt
3. Gala S Integrated Mission Group in Security (IMG-S); Online: http://www.nato.int/nato_static/assets/pdf/2013_10_pdf/20131125_7_Gala_SPS_CBRN_WS_2013.pdf
4. Jacobsen S NATO's political-military view on CBRN defence. Online: http://www.nato.int/nato_static/assets/pdf/2013_10_pdf/20131125_3_Jacobsen_SPS_CBRN_WS_2013.pdf
5. EU CBRN Action Plan. Online: http://www.ebsaweb.eu/ebsa_media/ANNEXE1_COM_2009_273-p-1304.pdf

6. Mignone A The European union's chemical, biological, radiological and nuclear centers of excellence initiative, non-proliferation papers, no. 28, June 2013; Online: http://www.nonproliferation.eu/documents/nonproliferationpapers/aliciamignone51bb2aa04185e.pdf
7. Centres of Excellence, an initiative of European Union. Online: http://www.cbrn-coe.eu/Home.aspx
8. Sabol J et al. The role of international organizations in combating radiological terrorism, security revue, 21 Feb 2011. Online: http://www.securityrevue.com/article/2011/02/the-role-of-international-organizations-in-combatting-radiological-terrorism/
9. (2013) Strategy of the Czech Republic for the fight against terrorism (from 2013 onwards), Ministry of interior of the Czech Republic, Security Policy Department, Prague
10. Osvald V CBRN first responders training course, Workshop on CBRN Defence, Brussels, 22–24 Oct 2013

Chapter 16
Will the Implementation of Small Modular Reactors Affect Our Response Protocols to Nuclear and Radiological Threats?

James Tom Voss and Robert Goldstein

Abstract The US is on the brink of implementing Small Modular Reactors as electricity generation systems. Those SMRs have many differences from the conventional Nuclear Power Plants we are familiar with. Those differences will substantially change our response protocols to their nuclear and radiological threats. We must first establish what the differences are then prepare our response protocols. We may learn much from other nations and international organizations which have invested considerable effort in investigating the nuclear and radiological threats presented by the implementation of Small Modular Reactors.

16.1 Introduction

A reliable and convenient source of electric power is vital to the US. However, those sources of electric power become more attractive to terrorist groups when coupled with the threat of nuclear and radiological attacks. Terrorist groups are no longer satisfied with simply creating fear in those they attack. Causing great harm is now the major goal of terrorist groups. Small Modular Reactors may present some unique opportunities for the terrorist groups. One of our challenges is to establish what those unique opportunities are then to prepare our threat response protocols to counter any terrorist activities related to those Small Modular Reactors. We may learn much from other nations and international organizations which have invested considerable effort in investigating the nuclear and radiological threats presented by the implementation of Small Modular Reactors.

J.T. Voss (✉)
CEO, Voss Associates, Abiquiu, NM, USA
e-mail: jtvoss@cybermesa.com

R. Goldstein
CEO, US Nuclear Corp, Canoga Park, CA, USA
e-mail: rgoldsteinta@gmail.com

S. Apikyan, D. Diamond (eds.), *Nuclear Threats and Security Challenges*, NATO Science for Peace and Security Series B: Physics and Biophysics,
DOI 10.1007/978-94-017-9894-5_16

16.2 What Are Small Modular Reactors?

Small Modular Reactor under development have electrical outputs ranging from 10 to 640 MW(e). According to the classification adopted by the IAEA, small reactors are reactors with an equivalent electric power of less than 300 MW(e). Using that classification all of the US Navy nuclear reactors could be classified as Small Modular Reactors. There are currently more than 40 designs for Small Modular Reactors. The IAEA states there are 131 Small Modular Reactors operating at present with another 45 under construction. The US military operates more than 50 Small Nuclear Reactors mainly in the US Navy. The SMR refueling cycles as stated by the developers range from 12 months to as long as 7–8 years. One SMR design has NO refueling cycle, the SMR operates for up to 30 years and is then retired, but the fuel is to be enriched to 16.3 %. This higher enrichment adds another attractive feature for any terrorist group. Many SMRs are operating currently in nuclear navies around the world and in some commercial applications. In the US the main developers are stating 45–180 MW(e) output [3, 4].

Global SMR Development – IAEA – Development Status of Small and Modular Reactors [1]

	mPower NuScale W-SMR Hi-SMUR	B&W received US-DOE funding for **mPower** design. The total funding is 452M$/5 years for 2 out of 4 competing iPWR based-SMRs. Some have utilities to deploy in specific sites. US-DOE also announced the second round of SMR funding in March 2013.
	SMART	On 4 July 2012, the Korean Nuclear Safety and Security Commission issued the Standard Design Approval for the 100 MWe **SMART** – the first iPWR received certification.
	KLT-40s SVBR-100 BREST-300 SHELF	Construction of 2 modules of barge-mounted **KLT-40s** near completion; Lead Bismuth cooled **SVBR-100** & Lead-cooled **BREST-300** to deploy by 2018, **SHELF** seabed-based conceptual design
	Flexblue	DCNS originated **Flexblue** capsule, 160 MWe, 60-100m seabed-moored, 5-15 km from the coast, off-shore and local control rooms
	CAREM-25	Site excavation for **CAREM-25** was started in September 2011; construction started in 2012
	4S	Toshiba had promoted the **4S** for a design certification with the US NRC for application in Alaska and newcomer countries.
	PFBR PHWRs: 220, 540 & 700, AHWR300-LEU	The **Prototype FBR** ready for commissioning and start-up test. 4 units of **PHWR-700** under construction, 4 more units to follow. **AHWR300-LEU** at final detailed design stage and ready for construction. 3
	CEFR HTR-PM ACP-100	2 modules of HTR-PM under construction; CNNC developing ACP-100 which will be deployed by 2018
	IRIS	Politecnico di Milano (POLIMI) and universities in Croatia & Japan are continuing the development of IRIS design - previously lead by the Westinghouse Consortium

Reactors Under Construction in the SMR Category – IAEA

Country	Reactor Model	Output (MWe)	Designer	Number of units	Site, Plant ID, and unit #	Commercial Start
Argentina	CAREM-25 (a prototype)	27	CNEA	1	CAREM-25	2017 ~ 2018
China	HTR-PM (GCR) (a prototype)	200	Tsinghua Univ./Harbin	1 (2 modules)	Shidaowan unit 1	2017 ~ 2018
India	PHWR 700	640	NPCIL	2	Kakrapar 3 and 4	6/2015 and 12/2015
	PHWR 700	640	NPCIL	2	Rajashtan units 7 and 8	6/2016 and 12/2016
	PFBR 500 (LMFBR)	500	IGCAR	1	PFBR Kalpakkam	2015
Pakistan	CNP-300	300	CNNC - China	2	Chasnupp 3 and 4	12/2016
Russian Federation	KLT-40S (ship-borne)	30	OKBM Afrikantov	2	Akademik Lomonosov	2012

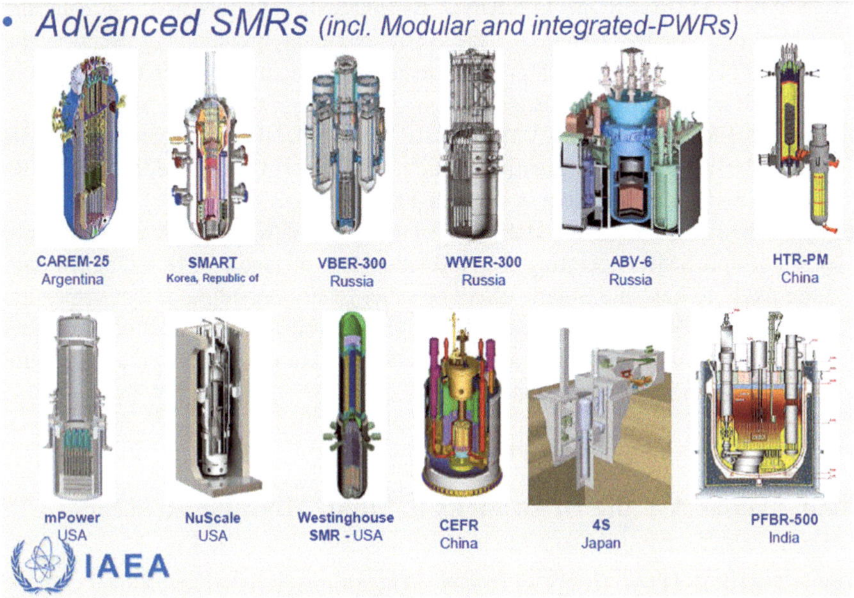

16.3 What Are the Advantages of Small Modular Reactors?

(a) LOWER INITIAL COST – The capital investment for a single SMR is much less than for a conventional nuclear power plant. To the electric utilities and their investors this means much less capital at risk by building SMRs instead of conventional nuclear power plants.

(b) SIMPLER ENVIRONMENTAL IMPACT STATEMENT – The physical footprint of a single SMR is much smaller than for a conventional nuclear power plant and the amount of radioactive material at risk is also much less.

(c) EASIER PUBLIC ACCEPTANCE – LOWER VISIBILITY – Small Modular Reactors have a much shorter construction time due to their smaller size and modular construction. To the public the construction of a Small Modular Reactor would appear to be less intrusive than the construction of a shopping mall.

(d) SERVICE REMOTE COMMUNITIES WITHOUT LONG DISTANCE TRANSMISSION LINES – Small Modular Reactors lend themselves to installation in remote areas that are not served by the electric grid (or not served well by the electric grid). Several proposals for the construction of SMRs advertise this feature as a selling point for SMRs.

(e) MODULAR CONSTRUCTION – SEMI ASSEMBLY LINE MANUFACTURING – SMRs with their modular construction can have all major components assembled off-site then transported to the operational site for final assembly. This modular construction method makes it appear to the public that the SMR almost magically appears.

(f) SHORT TIME FROM GROUND BREAKING TO ONLINE – With the simplified control systems of the SMR and their modular construction the period of time from final approval to actually having the SMR on the grid could be well less than 12 months.

(g) SIMPLER NRC APPROVAL AFTER FIRST UNIT – Keeping the SMR modules identical will greatly simplify NRC approvals for additional SMRs either at the same location or separate locations.

(h) SMALLER IMPACT IN THE EVENT OF AN INCIDENT – Even in the case of multiple SMRs operating at an individual site, no single SMR incident would be likely to have a dramatic effect on other SMRs operating at that same site. A single SMR has much less radioactive material at risk than a conventional nuclear reactor. Additionally the SMR designs are more robust than currently operating nuclear power plants.

16.4 What Are the Drawbacks to Small Modular Reactors?

(a) HIGHER INITIAL COST PER MW – The estimated initial investment cost per MW is perhaps 50 % higher than for a conventional nuclear reactor.

(b) MORE PERSONNEL PER MW – The earlier statements about SMRs operating without human intervention was obviously an exaggeration. The number of personnel per MW installed will always be higher for an SMR than for a conventional nuclear reactor. Installing multiple SMRs at a single location will somewhat reduce the number of personnel per MW but will not reduce to the number of personnel required for a conventional nuclear reactor.

(c) HIGHER INSTRUMENTATION COST PER MW – The number of process instruments per MW installed will always be higher for an SMR than for a conventional nuclear reactor. Installing multiple SMRs at asingle location will somewhat reduce the number of process instruments per MW but will not reduce to the number of process instruments required for a conventional nuclear reactor.

(d) HIGHER OPERATION COST PER MW – SMRs cannot equal the operational costs per MW of conventional nuclear reactors due to the economy of scale. Conventional nuclear reactors have an enviable record of production capacity. Extended operating time between outages and much shortened outages have pushed all the US commercial nuclear reactors into the mid to high 90 % capacity factors.

(e) PROTOTYPE STAGE ONLY IN THE US (except for our nuclear navy and commercial nuclear ships and barges) – Until the first SMR goes operational in the US they have to be considered prototypes.

(f) NO PROVEN SAFETY RECORD (except for our nuclear navy and commercial nuclear ships and barges) – Until SMRs become operational we cannot determine their safety record.

(g) NO NRC APPROVAL CRITERIA ESTABLISHED – Until an application for construction has been submitted to the NRC there cannot be an approval criteria established by the NRC. In addition the NRC criteria for required physical security cannot be established until a detailed application for construction has be submitted.

(h) NO APPLICATION FOR LICENSE – The current NRC estimate is that applications for license for SMRs may be received by late 2015.

16.5 Where Are Small Modular Reactors Currently Operating? [2]

Many SMRs are operating currently in nuclear navies around the world and in some commercial applications such as nuclear merchant vessels and power barges. Argentina, China, India, Japan, Korea, Pakistan, the Russian Federation, South Africa, and the US are progressing in the implementation of SMRs. Remote locations such as Alaska, the Antarctica, parts of Africa, Russia, China, India, South America, and even in less populated areas of North America are suggested locations for the implementation of SMRs [5].

Research in SMRs is ongoing in many nations around the world. One interesting experiment is the Los Alamos National Laboratory – Experimental Demonstration of a Heat Pipe/Stirling Engine Nuclear Reactor. The reactor was operated at a maximum 1,000 W(t) and during a separate test on electric generation produced 50 W(e) (but not at the 1,000 W(t) output).

Los Alamos National Laboratory – LA-UR-13-23137

Experimental Demonstration of a Heat Pipe/Stirling Engine Nuclear Reactor

(DUFF) experiment was devised as a proof-of-concept test for the aforementioned small space reactor.

The test was conducted in September of 2012 and was successful on many levels, but most importantly it produced electricity using nuclear fission, heat pipes, and Stirling engines. DUFF demonstrated the first Stirling engine to be powered with fission energy, the first use of a heat pipe to transport heat from a reactor to a power conversion system, and the first nuclear-powered test of a space reactor technology in over 40 years. One of the biggest benefits of the testing was that it provided valuable experimental data. The data demonstrated that the concept is viable and the physics is well characterized. The data also provided the ability to benchmark criticality and dynamic reactor modeling tools used to design the experiments.

II. DUFF EXPERMIMENTAL SETUP

The experiment to model the reactor consists of three main parts combined into a single experimental setup. The three parts of the experiment are:

1. The critical experiment assembly "Flattop"
2. A water based heat pipe
3. A Stirling Engine power conversion system

Fig. 2. Flattop critical experiment assembly with core

16.6 What Are World-Wide Plans for Small Modular Reactors?

The world-wide investment in the current generation SMRs being developed exceeds several billion dollars. With some SMRs nearing operational status future investment will continue to increase. Argentina, China, India, Japan, Korea, the Russian Federation, South Africa, and the US are all at an advance stage of having commercial SMRs in operation. The pros and cons of implementing SMRs seem to lead towards having small numbers of units in remote locations not served by large electric generating systems.

16.7 How Will We Maintain Security of Nuclear Materials?

(a) Will Small Modular Reactors Require More Fuel Shipments per MW – Since SMRs use fewer fuel bundles than conventional nuclear reactors the number of fuel bundles per MW installed will be higher for SMRs. Those increased numbers of fuel shipments mean we will have more shipments to monitor and protect. We will need more personnel than for conventional nuclear reactors. The same logic applies to any shipment of expended nuclear fuel.
(b) Will Remote Locations of Small Modular Reactors make them more attractive targets – With remote locations of SMRs there will be a longer lag time in receiving off-site assistance in the event of terrorist activity. The NRC requirements for SMR security needs to address this concern.

(c) Will higher enrichment in some Small Modular Reactor designs make them a more attractive target – The higher enrichment of fuel in some SMR designs will certainly be more attractive to terrorists. More radioactive material in a smaller package will be a point the terrorists will not miss.

16.8 Conclusion: Will the Implementation of Small Modular Reactors Affect Our Response Protocols to Nuclear and Radiological Threats?

(a) More nuclear reactors per MW installed.
(b) More Fuel Shipments per MW installed including spent fuel shipments.
(c) Remote Locations of Small Modular Reactors make them easier targets.
(d) With remote locations of SMRs there will be a longer lag time in receiving off-site assistance in the event of terrorist activity.
(e) Higher enrichment in some Small Modular Reactor designs make them a more attractive target.

There will be more fuel shipments per MW installed, there will be remote locations with less access to rapid response teams, higher enrichment fuel will be more attractive to terrorist groups. These lead to the need for more response members including more rapid response. Addressing the potential additional radiological hazard of higher enriched fuel must be considered.

We will need to improve our response protocols. First we must analyze the changes in the nature of the threats due to the implementation of Small Modular Reactors. There are no simple solutions but there are opportunities to determine what those solutions are.

References

1. Hadid Subki M Technical Lead for SMR Technology Development. IAEA - development status of small and modular reactors. Nuclear Power Technology Development Section Division of Nuclear Power, Department of Nuclear Energy presented at the 24th meeting of the IAEA TWG – NPPIC
2. World Nuclear Association – small nuclear power reactors. www.world-nuclear.org
3. US Department of Energy – small modular nuclear reactors www.energy.gov/ne/nuclear-reactor-technologies/small-modular-nuclear-reactors
4. Voss JT President of the Power Reactor Section of the HPS. Health Physics Society – small modular reactors. Presented at the July 2014 meeting of the Health Physics Society
5. Poston, David I, McClure, Patrick R, Dixon, David D, Gibson, Marc A, Mason, Lee S Experimental demonstration of a heat pipe/stirling engine nuclear reactor. Los Alamos National Laboratory – LA-UR-13-23137.

Part III
Advanced Technologies for Nuclear Safety and Security

Chapter 17
Advanced Concepts in Multi-dimensional Radiation Detection and Imaging

Kai Vetter, Dan Chivers, and Brian Quiter

Abstract Advanced concepts in radiation detection and imaging significantly enhance the capabilities relevant for nuclear security and safety as well as in prevention and in response to nuclear and radiological attack. More recent developments in combining radiological and nuclear detection concepts with complementary sensor data and information provide yet further improved capabilities in these areas as well as in risk management and mitigation. We briefly discuss some of the new concepts and technologies that are being developed and implemented in the Berkeley Applied Nuclear Physics program. They range from micrometer resolution scale instruments that enable new means in detecting and reconstructing gamma rays to meter-scale instruments necessary to enable standoff detection capabilities. Complementary to that, contextual and environmental data are being measured and correlated with nuclear signatures and backgrounds to increase the ability to detect weak sources in the midst of spatially and temporally varying backgrounds. The concept of three-dimensional, volumetric imaging will be described as well the concept of the Nuclear Street View, both related concepts relevant for the detection and characterization of nuclear materials and associated activities. Finally, the impact of these technologies in the effective assessment of structures and radiation after a radiological or nuclear event will be discussed.

17.1 Introduction and Context

The ability to detect radiation emanating from atomic nuclei or nuclear processes has applications ranging from basic research to medicine and security. Some of the objectives in these diverse applications are similar as the goal is to detect or estimate weak specific nuclear signatures in the midst of background or noise.

K. Vetter (✉) • D. Chivers • B. Quiter
Lawrence Berkeley National Laboratory, Berkeley, CA, USA

K. Vetter
University of California, Berkeley, CA, USA
e-mail: kvetter@berkeley.edu

S. Apikyan, D. Diamond (eds.), *Nuclear Threats and Security Challenges*,
NATO Science for Peace and Security Series B: Physics and Biophysics,
DOI 10.1007/978-94-017-9894-5_17

Conceptually, we can differentiate between detection and estimation. In detection, a binary decision matrix can be defined capturing all possible outcomes in being right or wrong about a detection or non-detection. In estimation, a non-binary function is to be measured, either as a quantity or shape or any other property of interest. Simple radiation counters, more advanced spectrometers, or imagers can be operated for both purposes, just for detection or for estimation. In any domain and implementation, the challenge is to discriminate between relevant and non-relevant information and to distinguish the signal of interest and background or the signal and noise with high sensitivity and specificity.

Particularly after the events of 9/11/2001 significant resources have been invested to enhance the nuclear detection capabilities nationally in the U.S. and globally. Radiation sensors are now abundantly distributed at border crossings and in cities, either deployed stationary or on mobile platforms or personnel. The main goal of these systems is the early detection of the proliferation of radiological or nuclear materials that could be used in a terrorist attack. The detection includes the interdiction of materials during transport or while being stored temporarily. It also includes the detection specifically of undeclared activities resulting in the diversion of materials in nuclear or radiological facilities. More recently the importance of background radiation has been recognized that needs to be taken into account in detecting and estimating nuclear materials and associated activities. While it might not possible to influence the signal of interest, e.g. the amount and type of radiological material and shielding, understanding and estimating the background effectively will increase the detection sensitivity. One of the often-used metrics for detection sensitivity is the Receiver Operator Characteristic (ROC) curve which shows the detection probability (PD) as a function of the false-alarm rate (FAR) for a specific detection modality and a given detection scenario. Both, the PD and FAR are critically important in operating detection systems as a high PD ensures the detection of even weak signals and a low FAR ensures a minimum impact of nuisance alarms that need to be addressed. Keeping the FAR low is challenging for hand-held sensors as well as far large systems for standoff detection. The challenge in keeping the PD high and the FAR low for weak signals, particularly when gamma rays are used, is compounded by the fact that radiation backgrounds change spatially and temporarily. Finding ways to better estimate or to remove background from the signal of interest is still a very active field of research. In this paper, we discuss complementary approaches in this research, ranging from advanced concepts in gamma-ray imaging to multi-sensor fusion to extensive background measurements, the latter to better understand the variations of radiation backgrounds in real-world environments.

17.2 Advanced Detection and Estimation Approaches

Most of the detection systems deployed today are simple counting instruments ranging from compact active dosimeters to large plastic portal monitors. In order to minimize or to mitigate the low specificity, spectroscopic gamma-ray detectors are used that are able to discriminate between non-threat and threat sources. Non-threat sources can be naturally occurring materials or medical or industrial sources. Finding ways to enhance the understanding and estimating background variations can enhance the ability to unambiguously detect threat sources. Complementary to spectroscopy, imaging concepts can be used to discriminate point-like threat sources over extended backgrounds or by correlating spectral signatures with specific locations. A gamma ray whose energy is measured is worth more than just the registration of its existence. The incident direction and the location of the source provide additional value. However, both, the spectroscopy and imaging are associated with decreased efficiency as not all gamma rays deposit their full energy in a detector and only a fraction of the incident gamma rays can be imaged. The gain in specificity due to spectroscopy, imaging, or a combination of both needs to balanced with the loss in efficiency and the additional costs and complexity of the deployed systems.

In applications such as nuclear safeguards or treaty verification, the imaging capability could be worth more than in detection and non-proliferation. In safeguards, nuclear processes and designs need to be verified, undeclared activities identified, and materials need to be accounted for. In radiological or nuclear storage or waste facilities or in circumstances when access is limited, when materials need to be assessed, verified, or detected, the location and extension of sources are of interest. Or in general, when complex radiation fields need to be analyzed. These objectives can benefit from gamma-ray imaging, particularly when the objects containing the nuclear and radiological materials are known or can be measured simultaneously and combined with the gamma-ray image information.

In the following, we discuss a new concept in gamma-ray imaging that can potentially enable more specific and efficient gamma-ray imaging and the concept of three-dimensional volumetric imaging reflecting the ability to produce or use three-dimensional models for gamma-ray reconstruction in three dimensions. While the first concept is still in its early phase of development, the 3D volumetric imaging is much closer for deployment. As a matter of fact, we developed the idea and had the first conceptual demonstration about 7 years ago already [1, 2].

17.2.1 Electron-Tracking Based Compton Imaging

Conventional means of imaging individual gamma rays as widely used in nuclear medicine rely on collimators to determine the incident angle or the so-called line-of-response. They work satisfactorily at gamma-ray energies of 140 keV or up to 400 keV, at higher energies with significantly reduced efficiency due to the limited attenuation of gamma rays in the collimator. Coded aperture based imaging provides higher efficiency, up to 50 %, however, are limited to simple source configurations. In addition, passive collimators and apertures are heavy, preventing easy portability.

An alternative approach for gamma-ray imaging relies on the ability to measure energies and three-dimensional positions of individual gamma rays

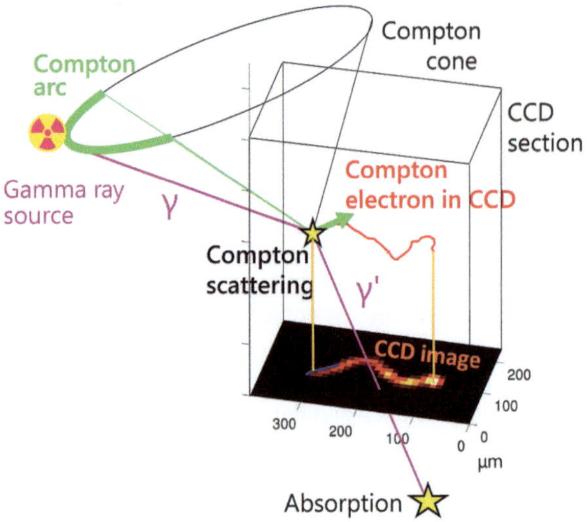

Fig. 17.1 Illustration of electron-tracking based Compton imaging. Without the knowledge of the electron scatter direction, the direction of the incident gamma can only be restricted to a cone surface. By measuring the initial electron trajectory the cone can be reduced to an arc reducing the background in the reconstructed gamma-ray image. The resolution, i.e. the extension of the arc depends on the resolution of the electron track reconstruction. The intrinsic width of the cone and arc depends on the energy and position resolution for individual gamma-ray interaction of the instrument and ultimately, the uncertainty due to the intrinsic momentum of the Compton-scattered electron

interactions [3, 4]. These can be used to reconstruct the direction of the incident gamma ray to a surface of a cone. As shown in Fig. 17.1, the two first interactions provide the symmetry axis of the cone, the opening angle is determined by the energy of the first interaction and the total energy of the incident gamma ray. Since this concept relies on the Compton scattering process as its first interaction, it is called Compton imaging. The advantage of Compton imaging is the fact that no heavy collimator is needed limiting the field-of-view and efficiency and is characterized by low weight, specifically for high gamma-ray energies. The drawback is the requirement of position-sensitive detectors and the Compton scattering process. Because of the latter, Compton imaging does not work or has limited imaging performance at low energies, typically below 150 keV. Another disadvantage of Compton imaging is the ability to only reconstruct a cone for each event, reflecting the symmetry of the Compton scattering process. This symmetry can be broken if one can measure the initial direction of the Compton-scattered electron. This so-called electron-tracking based Compton imaging allows us to reconstruct an arc instead of a full cone on event-by-event basis, as shown in Fig. 17.1. Figure 17.2 illustrates conceptually the gain one can obtain by employing electron tracking as compared with conventional Compton imaging.

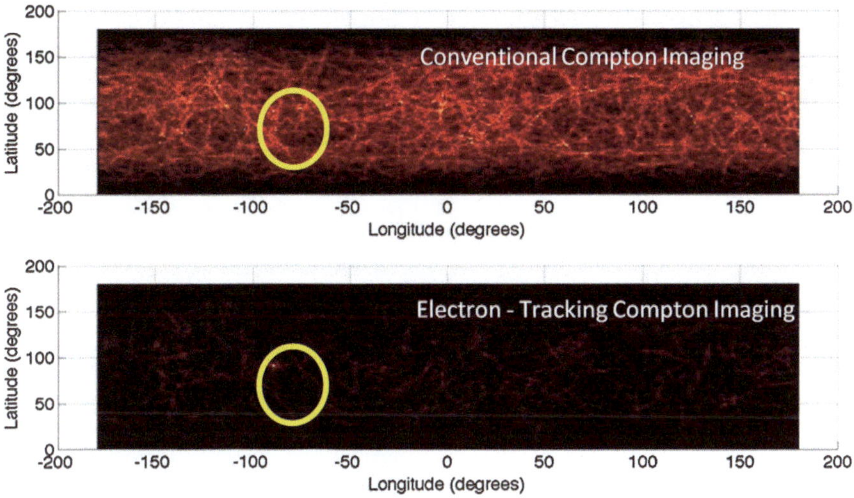

Fig. 17.2 Illustration of potential gain by employing electron-tracking based Compton imaging in comparison with conventional Compton imaging. Simple cone backprojection is used assuming an angular electron track reconstruction resolution of 30

We were able, for the first time, to demonstrate electron-tracking based Compton-imaging in a solid-state detector [5, 6]. While this concept has been shown before in gas-based instruments, these detectors are characterized by very low interaction efficiency and low energy resolution, significantly limiting the achievable gamma-ray imaging sensitivity [7]. The scientific CCDs we are using are fully depleted Si devices with a thickness of 650 um and a pixel pitch of 10.5 μm [8]. These performance characteristics are well suited for tracking electrons above about 150 keV. We have demonstrated the ability to reconstruct gamma rays with a arc opening angle of about 30–60°, depending on incident angle and deposited energy [9].

While the accurate and efficient measurement of the initial direction of the scattered electron represents a significant challenge for the detection instrument, the conversion of the measured energy and position signals into an energy and direction of the incident gamma rays requires sophisticated and newly developed algorithms. For example, we have developed algorithms to extract the initial three-dimensional scatter direction of the electron by only observing the two-dimensional energy loss distributions in the used CCD [10]. Furthermore, we have developed a new algorithm that allows us to analytically reconstruct the incident gamma-ray energy and angle by just using the initial electron scatter direction and the energy deposited by the Compton scattered electron, i.e. by only measuring the partial energy of the incident gamma ray [11]. This realization potentially allows us to reconsider inherently inefficient gas detectors, as only one interaction is required. Figure 17.3 illustrates the results of the reconstruction of a Cs-137 source characterized by a gamma-ray energy of 662 keV.

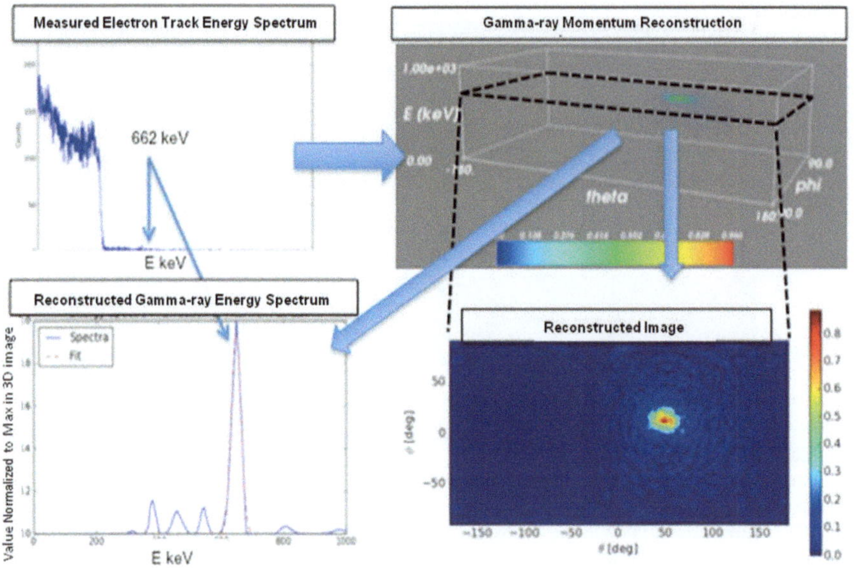

Fig. 17.3 Reconstruction of gamma-ray momentum, i.e. energy and incident direction by using the measured initial electron scatter direction and electron energy in a CCD system. *Top left* is the measured electron energy in the CCD reflecting the energy deposited by the Compton scattered electron. Note that the full energy of the gamma ray is not visible as only the energy of the Compton-scattered electron is deposited. *Top right*: Reconstructed gamma-ray momentum. *Bottom left*: Projection of momentum on energy axis illustrating the reconstruction of the incident, full gamma-ray energy. *Bottom right*: Reconstructed incident direction of gamma rays

17.2.2 3D Volumetric Imaging

Complementary to μm-resolution and small-scale instruments used to develop a new generation of gamma-ray imaging instruments as discussed above, we are pursuing the development of medium-sized and hand-portable gamma-ray detection and imaging systems in combination with contextual sensors. These combined 3D volumetric imaging systems enable the simultaneous reconstruction of the 3D environment or scene and the fusion of gamma-ray images with this 3D world. This 3D fusion goes beyond the already developed overlay of visual images and gamma-ray images in two-dimension by fully integrating multi-sensor information in three dimensions. The advantages are the significantly increased efficiency, accuracy, and contrast in the measurement of the gamma-ray information, the possibility for quantification of gamma-ray sources, and the increased speed in the gamma-ray reconstruction. As development platform, we are using the second generation Compact Compton Imager (CCI-2) and the High-Efficiency Multimode Imager (HEMI) [12–14]. Both systems are shown in Fig. 17.4.

CCI-2 consists of two high-purity Ge (HPGe) detectors implemented in a double-sided strip detector (DSSD) configuration. Each detector has a thickness of 15 mm

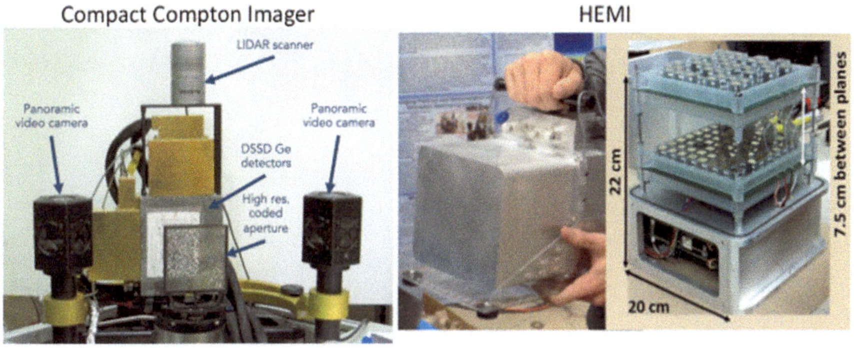

Fig. 17.4 *Left*: Second generation Compact Compton Imager consisting of 2 DSSD HPGe detectors mounted in a cryostat with contextual sensors, two panoramic video cameras and a Lidar scanner. *Right*: The hand-portable High-Efficiency Multimode Imager consisting of arrays of CPG CZT detectors to enable coded aperture and Compton imaging, simultaneously

and 38 orthogonal strips with a strip pitch of 2 mm. Digital signal processing is used to obtain the position of individual gamma-ray interactions to about 1 mm in all three dimensions. This position resolution combined with the energy resolution of about 2 keV for each strip provides an angular resolution of about 4° when operated in Compton imaging mode. The angular resolution depends on the energy and data cuts implemented reflecting the trade-off between efficiency and angular resolution. The system is mounted on a cart providing support for the fully digital data acquisition system to read out the 152 strips, the power supplies, and the liquid nitrogen dewar to provide the necessary cooling for the HPGe detectors. While this system is too heavy for easy deployment and represents an efficient platform for the development of detection and data fusion concepts. A more compact and hand-portable, DSSD HPGe based system is commercially available, but with reduced position resolution [15]. HEMI consists of 96 CdZnTe (CZT) detector implemented in a coplanar grid (CPG) configuration [16]. The individual, 1 cm^3 big CZT detectors are arranged in two parallel planes with 32 detectors in the front plane and a fully populated plane of 64 detectors arrange in a 8×8 array. The front plane of active detectors serves as coded aperture for low gamma-ray energies and as scatter plane for Compton imaging for higher gamma-ray energies. The total active detector mass is 1.2 lbs with the whole instrument weighing about 8 lbs and is therefore easily portable. In comparison to the DSSD HPGe the energy resolution is degraded and the position resolution is limited to 1 cm^3 resulting in an angular resolution of about 10°at 662 keV. Both systems have been combined with contextual sensors such as high-resolution Ladybug video cameras and 32 beam rotating Lidar systems or Microsoft Kinect© sensors.

Figure 17.5 illustrates the reconstruction and fusion of objects and images in 3D. By moving the cart or walking with HEMI through the scene, it is possible to reconstruct a point cloud representing surfaces of objects in 3D and to use these point clouds to increment the gamma-ray image.

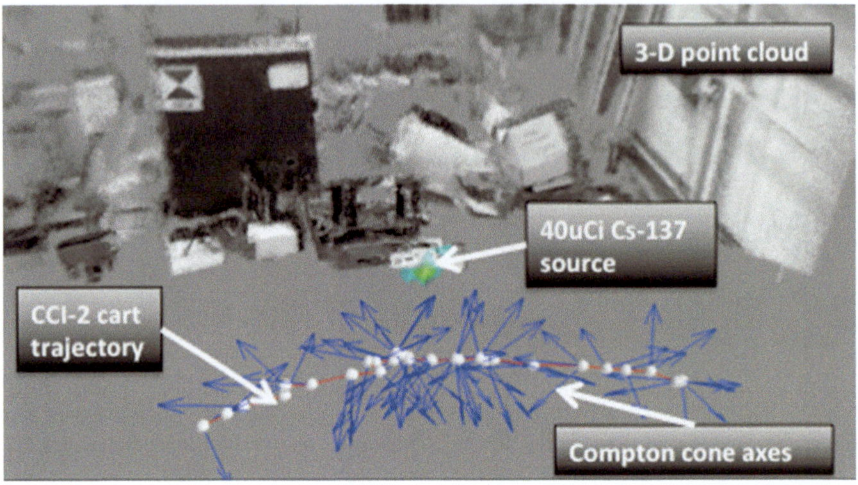

Fig. 17.5 Illustration of 3D volumetric imaging. 3D point clouds are registered using a Microsoft Kinect sensor mounted to the CCI2 gamma-ray imaging systems. By moving the cart through the cluttered scene, the 3D point cloud is obtained and subsequently used to reconstruct the gamma-ray image in 3D. The *red line* indicates the trajectory of the instrument, the *blue arrows* represent the gamma-ray scatter directions. These events have been used to reconstruct the location of the Cs-137 source. The total acquisition time was less than 90 sec. while the cart was moved over a distance of about 12 ft

In collaboration with JAEA, HEMI has also been deployed to Fukushima to evaluate contamination mapping capabilities when mounted on a RMAX unmanned and remotely controlled helicopter. The dimensions and the weight of HEMI are well suited for these demonstrations. The HEMI instrument itself was packaged in an environmental box that allowed an operation completely independently from the helicopter operation. The whole package included a video camera, a GPS/IMU system, thermal stabilization, data storage, batteries, a control unit, and HEMI. The video information is used to reconstruct three-dimensional surfaces on the ground stereoscopically. These surfaces are then used to project the gamma-ray images onto as shown in Fig. 17.6. Although these 3D surfaces improve the reconstruction, it still provides limited reconstruction capabilities. This is due to the fact that the surfaces are typically the surfaces of vegetation and not the source of the radiation which is located beneath the top surfaces, e.g. in the soil or the whole plant. More work remains to be done to increase the quantitative reconstruction capabilities in the complex environments of Fukushima.

17.2.3 Large-Scale Detection Systems

Funded by the Domestic Nuclear Detection Office (DNDO), we have been operating the truck-based Radiological Multi-sensor Acquisition Platform RadMAP in the

Fig. 17.6 *Left*: HEMI as mounted in an environmental box and on a RMAX helicopter in the evacuated area of the Fukushima Prefecture. *Right*: Compton image reconstruction of Cs onto 3D surfaces obtained by using the onboard video camera

Bay Area for the last 3 years. We are pursuing three main objectives with this system: (1) Systematic mapping of radiation backgrounds; (2) Evaluation of detection concepts in realistic environments; (3) Development of the Nuclear Street View [17]. A systematic mapping of radiation backgrounds in a range of environments with their temporally and temporary variations is important to evaluate and develop improved detection concepts. When searching for weak sources, the ability to estimate the background is critical. For detection, not only the sensitivity to detect a source is important but also the specificity in avoiding false alarms. Conventionally, predominantly the statistical noise terms were considered when evaluating detection performance, neglecting systematic variations in backgrounds that are found in realistic environment, particularly when operating on a mobile platform. The new concept of the Nuclear Street View is an extension of the 3D volumetric imaging and the 3D Google Streetview©. It is based on the ability to reconstruct buildings in 3D as the conventional Streetview, but in addition, it is able to classify objects such as buildings and can associate gamma-ray emission spectra to these objects in 3D. It will be discussed further below in more detail.

RadMAP as shown in Fig. 17.7 is a truck-mounted platform that contains a range of nuclear radiation detectors as well as contextual and environmental sensors [17]. The nuclear radiation instruments consist of up to 24 large-volume HPGe gamma-ray spectrometers, 100 $4'' \times 4'' \times 2''$ NaI(Tl) gamma-ray detectors arranged in a 10×10 array behind a passive coded mask, and 16 large-volume liquid scintillation detectors for neutrons. The HPGe detectors provide excellent spectroscopic detection capabilities and the coded aperture system provides gamma-ray imaging capabilities. The NaI(Tl) detectors can also being used for detection. Two Ladybug video cameras and two rotating Lidar systems are mounted in the front to obtain a near-complete coverage of the environment. A digital weather station is mounted on the top to provide meteorological data. In addition, two hyperspectral cameras

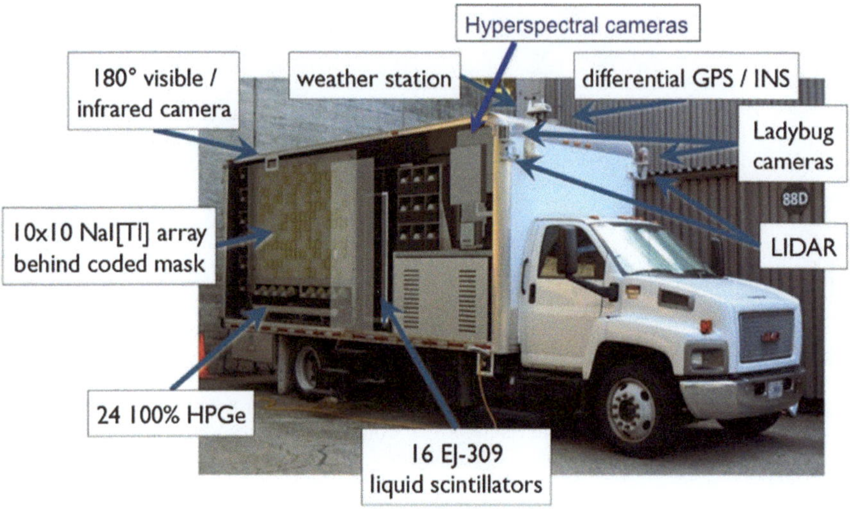

Fig. 17.7 The Radiological Multi-sensor Acquisition Platform RadMAP with its nuclear radiation detectors as well as contextual and environmental sensors

are mounted on the passenger side to analyze objects with regard to their emission spectra between 400 and 1700 nm, i.e. the visual and near-infrared spectrum.

With this system, large sets of background data and contextual information are observed and stored at the NERSC computing facility at LBNL. Figure 17.8 shows a map of routes driven in the areas of Oakland and Alameda. The colors represent the intensities of gamma rays associated with the decays of the main primordial background sources, K-40, Th-232, and U-238. The insert shows the three intensities observed close to a stainless steel bridge in Alameda [18].

Of particular interest in using RadMAP in this configuration is not only the ability to obtain extensive sets of gamma-ray and neutron backgrounds with contextual information, it also enables important comparisons of detection modalities [17]. Using the concept of source injection, it is possible to create ROC curves for a range of gamma-ray sources and intensities, standoff distances, truck speeds, etc. and for each of the detection instruments on board. The spectroscopy of the HPGe detectors can be compared with the spectroscopy of NaI(Tl) detectors or their imaging capability when operated with the coded mask. Algorithms can be evaluated and optimized for all detection modalities and detector types. In addition, the number of detectors can be changed and its impact in detection can be studied. As an example, Fig. 17.9 shows ROC curves for a range of detector configurations and algorithms when searching for a 1 mCi Cs-137 source at 100 m distance and driving at 25 mph. The axes were scaled so that a Gaussian distribution will result in a straight line in the log-log presentation. We have performed extensive benchmark measurements and simulations and find excellent agreement in the full-energy

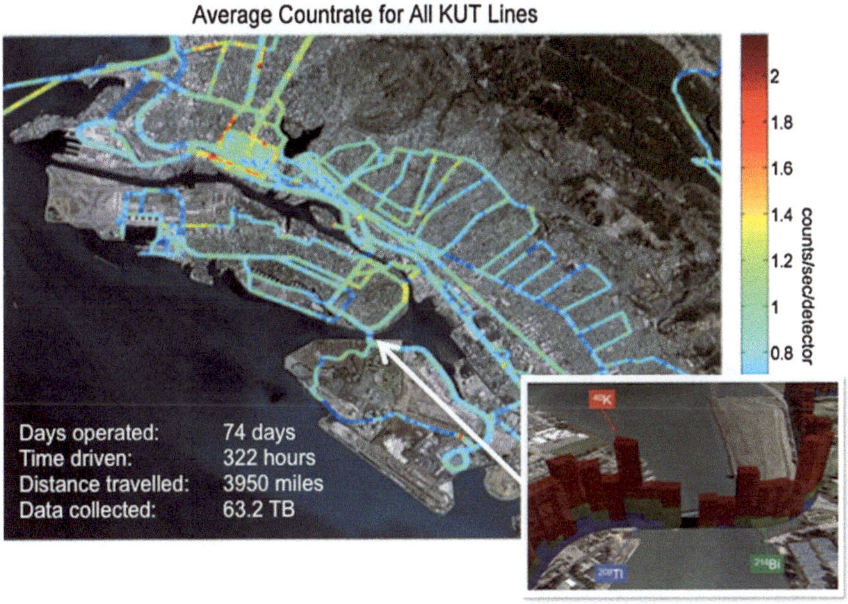

Fig. 17.8 Measured average gamma-ray count rates in HPGe detectors associated with K-40, U-238, and Th-232 (*KUT*). During 74 days of measurements almost 4,000 miles in the greater Bay Area were traveled. The highest count rate is about 2 cts per HPGe detector. The insert illustrates the differentiated count rates in K-40, U-238, and Th-232 and their variations in the vicinity of a stainless steel bridge

efficiencies of the HPGe and NaI(Tl) detectors [18]. We are currently developing and demonstrating new detection algorithms that make use of the whole spectra and contextual information to better estimate the background contribution to the total spectrum.

The large sets of data collected in the measurement campaigns are being stored, processed, annotated, and indexed using tools such as the FastBit and FastQuery developed at LBNL [19]. The goal is to make these data available to the user community. It also allows us to establish benchmarks for a wide range of detection hardware and software that is critical in guiding further developments and improvements. We have established the Berkeley Nuclear Data Cloud that allows registered users to access selectable data sets.

17.3 Nuclear Street View

As mentioned before, one of the goals of RadMAP is the development and demonstration of the Nuclear Street View. Figure 17.10 illustrates the concept. The Lidar and visual sensors are used to reconstruct surfaces in 3D, including the relative

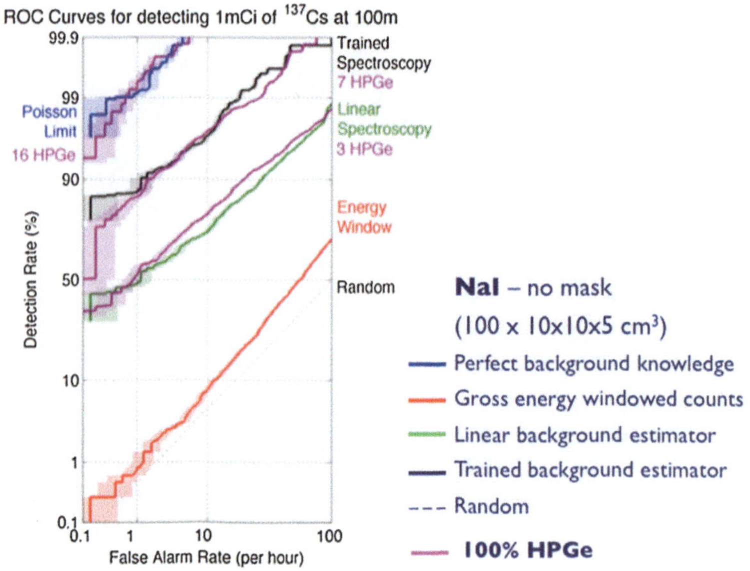

Fig. 17.9 Receiver Operator Characteristic (*ROC*) curves for detecting a 1 mCi Cs-137 source at 100 m standoff distance when driving with 25 mph. Detection is done by spectral analysis of the 100 NaI(Tl) and the indicated number of 100 % relative efficiency HPGe detectors mounted on the RadMAP platform

position and orientation of the truck within the reconstructed scenes. Semantic segmentation is being used to automatically classify objects such as building, vegetation, sidewalks, cars, persons, etc. The Nuclear Street View concept is able to associate spectral signatures to these specific objects. It is enabled by combining the semantic segmentation with the gamma-ray imaging data and the knowledge of the precise response of the instrument. The Nuclear Street View concept will not only allow to associate gamma-ray signatures with objects such as buildings, it can potentially enable the prediction of backgrounds given specific non-radiation signatures. Such signatures are currently being studied by using the hyperspectral cameras on board of RadMAP.

17.4 Summary and Outlook

Advances in radiation detection and sensor technologies along with fast signal processing capabilities and algorithms continue to provide new and improved capabilities in the detection and mapping of radiological and nuclear materials. New μm resolution radiation detectors create information that enables unprecedented

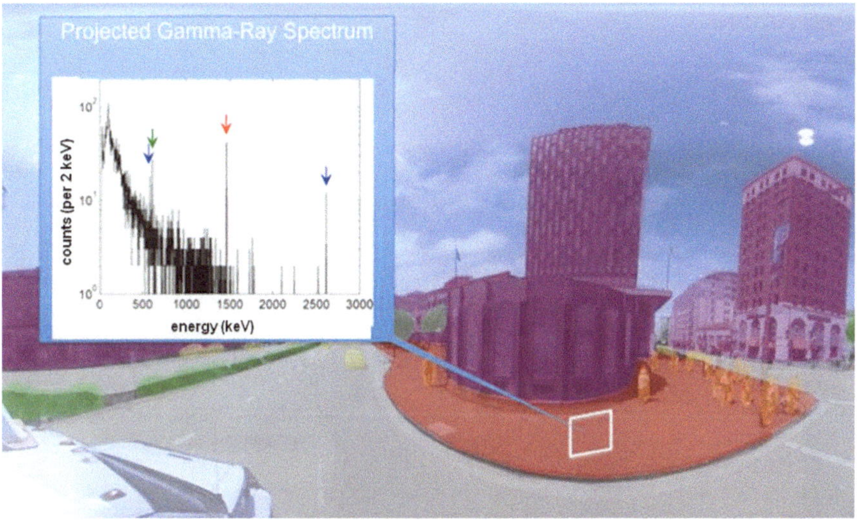

Fig. 17.10 Conceptual illustration of the Nuclear Street View. Semantic segmentation algorithms are applied to data collected with the video cameras and Lidar systems on board of RadMAP. Specific objects such as buildings, sidewalks, vegetation, cars, or people can be associated with gamma-ray emission signatures. Other classes can be identified, particularly when using different sensors such as the hyperspectral cameras that were recently mounted on RadMAP

capabilities in reconstructing the direction and energies of gamma rays. Hand-portable, high-energy resolution and position sensitive semiconductor detectors are now available that provide good spectroscopy and imaging performance. In combination with contextual sensors, gamma-rays image data can be fused with 3D scene data simultaneously and in near realtime. The use of contextual object data significantly increases the accuracy in the reconstruction of gamma-ray source location, the speed in the reconstruction, and ultimately enables the determination of the source strength. This 3D volumetric imaging can be applied inside nuclear processing or waste facilities or outdoors when searching for weak sources or mapping contamination in complex terrains. These new and advancing capabilities are and will continue to play a critical role in the prevention or in the response to radiological or nuclear events.

Acknowledgements The work presented here reflects contribution of many scientists and students, all members of the Berkeley Applied Nuclear Physics program. Specifically, I want to thank Mark Amman, Mark Bandstra, Andy Haefner, Paul Luke, and Lucian Mihailescu from LBNL and Tim Aucott, Brian Plimley, Ryan Pavlovsky, Joey Curtis, and Ross Barnowski from UC Berkeley. This work has been supported by the US Department of Homeland Security, Domestic Nuclear Detection Office, under competitively awarded contract 2011-DN-077-ARI049-03 as well as under Grant Award Number ECCS-1140069. In addition, it is based upon work supported by the Department of Energy National Nuclear Security Administration under Award Number(s) DE-NA0000979.

References

1. Mihailescu L, Vetter K, Burks M, Hull E, Craig W (2007) SPEIR: a Compton camera. Nucl Inst Methods A 570:89
2. Mihailescu L, Vetter K, Chivers DH (2009) Standoff 3D gamma-ray imaging. Nucl Sci IEEE Trans 56(2):479
3. Vetter K (2007) Recent developments in the fabrication and operation of germanium detectors. Annu Rev Nucl Part Phys 57:326
4. Vetter K, Burks M, Cork C, Cunningham M, Chivers D, Hull E, Krings T, Manini H, Mihailescu L, Nelson K, Protic D, Valentine J, Wright D (2007) High-sensitivity Compton imaging with position-sensitive Si and Ge detectors. Nucl Inst Methods A 579:363
5. Vetter K, Chivers D, Plimley B, Coffer A, Aucott T, Looker Q (2011) First demonstration of electron-tracking based Compton imaging in solid-state detectors. Nucl Inst Methods A 652:599
6. Chivers DH, Coffer A, Plimley B, Vetter K (2011) Impact of measuring electron tracks in high-resolution scientific charge-coupled devices within Compton imaging systems. Nucl Inst Methods A 654:244
7. Takada A et al (2005) Development of an advanced Compton camera with gaseous TPC and scintillator. Nucl Inst Methods A 546(1–2):258
8. Holland SE et al (2003) Fully depleted, back-illuminated charge-coupled devices fabricated on high-resistivity silicon. Electron Devices IEEE Trans 50(1):225
9. Plimley B, Chivers D, Coffer A, Aucott T, Wanga W, Vetter K (2011) Reconstruction of electron trajectories in high-resolution Si devices for advanced Compton imaging. Nucl Inst Methods A 652:595–598
10. Plimley, Member, Chivers D, Coffer A, Member, Vetter K (2013) Experimental benchmark of electron trajectory reconstruction algorithm for advanced Compton imaging. IEEE Trans Nucl Sci 60(3):2308
11. Haefner A, Gunter D, Plimley B, Pavlovsky R, Vetter K (2014) Gamma-ray momentum reconstruction from Compton electron trajectories by filtered back-projection. Appl Phys Lett 105:184101
12. Amman M, Luke PN, Lee JS, Mihailescu L, Vetter K, Zoglauer A, Wunderer CB, Galloway M, Boggs SE, Chen H, Marthandam P, Awadalla S, Taherion S, Bindley G (2009) Detector module development for the high efficiency multi-mode imager. IEEE Trans Nucl Sci: 981–985
13. Galloway ML et al. Status of the high efficiency multimode imager. Nuclear Science Symposium and Medical Imaging Conference (NSS/MIC), IEEE. 2011, p 1290
14. Galloway M et al (2011) Simulation and detector response for the high efficiency multimode imager. Nucl Inst Methods A 652(1):641
15. PHDS gamma-ray imaging detectors. http://www.phdsco.com/products/gegi/specs
16. Luke PN (1994) Single-polarity charge sensing in ionization detectors using coplanar electrodes. Appl Phys Lett 65(22):2884
17. Aucott T, Bandstra M, Negut V, Curtis J, Vetter K (2014) Effects of background on gamma-ray detection for mobile spectroscopy and imaging systems. IEEE Trans Nucl Sci 61(2):985
18. Aucott T, Bandstra M, Chivers D, Negut V, Cooper R, Vetter K (2013) Routine surveys for gamma-ray background characterization. IEEE Trans Nucl Sci 60(2):1147
19. Wu K, Otoo EJ, Shoshani A (2006) Optimizing bitmap indices with efficient compression. ACM Trans Database Syst 31(1):1. https://sdm.lbl.gov/fastbit/

Chapter 18
Dynamics of International Nuclear Safety: Post-Fukushima Developments in Regulatory Oversight and Filtered Vent Technology in Six Nuclear Countries

Kathleen Araújo

Abstract Nuclear safety has undergone intense scrutiny since the Fukushima Daiichi accident in 2011. Areas of particular consideration include the strength of regulatory entities and the use of filtered containment venting systems (FCVS). This assessment reviews the status of nuclear safety regulators with an emphasis on independence. The chapter also examines rules on FCVS and its utilization. Developments specific to China, France, India, Russia, South Korea, and the United States are the central focus. The chapter finds varying levels of autonomy exist among the nuclear safety regulatory entities. The assessment also shows that four of the six countries have not adopted filtering technology, despite its use for decades and related safety gains. Aspects for continued monitoring and study are discussed.

Abbreviations

AERB	Atomic Energy Regulatory Board
ASN	Autorité de sûreté nucléaire
BDBA	Beyond design basis accident
BWR	Boiling Water Reactor
CNS	Convention on Nuclear Safety
CSNI	Committee on Safety of Nuclear Installations

This chapter draws on research that was completed while working as a joint fellow in Managing the Atom, and the Science, Technology and Public Policy Program at the Harvard Kennedy School of Government, and subsequently.

K. Araújo (✉)
Stony Brook University, Stony Brook, NY, USA

Massachusetts Institute of Technology, Cambridge, MA, USA
e-mail: Kathleen.Araujo@stonybrook.edu

S. Apikyan, D. Diamond (eds.), *Nuclear Threats and Security Challenges*, NATO Science for Peace and Security Series B: Physics and Biophysics, DOI 10.1007/978-94-017-9894-5_18

EPR	European pressurized reactor
FBR	Fast breeder reactor
FCVS	Filtered containment venting systems
FYP	Five Year Plan
GAO	Government Accountability Office
IAEA	International Atomic Energy Agency
INPO	Institute of Nuclear Power Operations
IRRS	Integrated Regulatory Review Service
MEP	Ministry of Environmental Protection
MWe	Megawatts of electric output
NEA	Nuclear Energy Agency
NEI	Nuclear Energy Institute
NLB	Nuclear Law Bulletin
NNSA	National Nuclear Safety Administration
NPP	Nuclear power plant
NRC	Nuclear Regulatory Commission
NSSC	Nuclear Safety and Security Commission
OECD	Organization for Economic Cooperation and Development
PHWR	Pressurized Heavy Water Reactor
PWR	Pressurized Water Reactor
RBMK	Reaktor Bolshoy Moshchnosti Kanalnyy
TMI	Three Mile Island
UNSCEAR	United Nations Scientific Committee on Effects of Atomic Radiation
US	United States
VVER	Vodo-Vodyanoi Energetichesky Reaktor
WNA	World Nuclear Association

18.1 Introduction

The Fukushima accident in 2011 spurred a new round of thinking about how nuclear safety is defined and managed. The events which played out at the Daiichi nuclear plant, starting March 11, 2011, highlighted a need for improved planning and operational focus in the context of beyond design basis accidents (BDBAs). The specific circumstances of the Fukushima accident were triggered by a combination of natural disasters – a 9.0 magnitude earthquake and 46–50 ft tsunami – with emergency nuclear plant conditions escalating from flooding, a prolonged loss of power (conventional and back-up), severe core damage and melting of fuel, hydrogen explosions, and radiation releases.

In the aftermath of the accident, analyses have highlighted lessons relating to process, design, and institutions [1–5]. Among key insights were the following:

- Planning and risk assessments should account for severe flooding and seismicity, station black-outs, and losses of ultimate heat sinks. Preparedness must also factor for severe accidents affecting multiple reactors. This entails multi-unit interactions which, in turn, can impact the availability of resources and people as well as the operability of emergency centers.
- Regular updates are needed for assessments and methodologies to factor for changes in knowledge, conditions, and better practices.
- Boiling Water Reactors of the Mark I and Mark II design are particularly vulnerable during severe accident conditions, given their small containment volume and dependence on suppression pools that do not mitigate hydrogen.
- Venting should operate passively without power. In addition, filtered venting (if included or added) can provide a crucial line of defense for people and the environment from harmful radiation during extreme events.
- The independence of nuclear regulators can play a pivotal role in cross-sectoral vigilance and a culture of safety. Regulators should be empowered with clear authority and ample capacity. Industry promotion should also be separate from oversight.
- The command-and-control process for emergency response should be well understood, particularly with respect to decision-making and clarity of roles. Processes and modalities for emergency communications should also be regularly evaluated.

This chapter briefly discusses the current state of global nuclear energy generation. It then examines developments relating to safety regulatory bodies and filtering technology in major nuclear states. The chapter closes with a discussion of areas to watch and questions for further study.

18.2 Global Nuclear Development

Worldwide, commercial nuclear generation, as of November 2014, could be found in 31 countries with a total of 438 nuclear reactors in operation, equaling 375,504 MWe of installed capacity [6]. Plants are under construction in new nuclearizing countries, like Belarus and the United Arab Emirates. The majority of new construction, however, is occurring in China and Russia (Ibid).

Among established nuclear states, the United States, France, Russia, South Korea, China, and India are some of the largest, collectively representing 62 % of total net electrical capacity and 59 % of nuclear power plants (NPPs) globally [6].[1] These six countries are the basis for the current comparative review (Figs. 18.1 and 18.2).

[1] Japan is also a major nuclear state with an energy policy which includes nuclear generation. However, the country's nuclear program remains on hold, as it recovers following the accident.

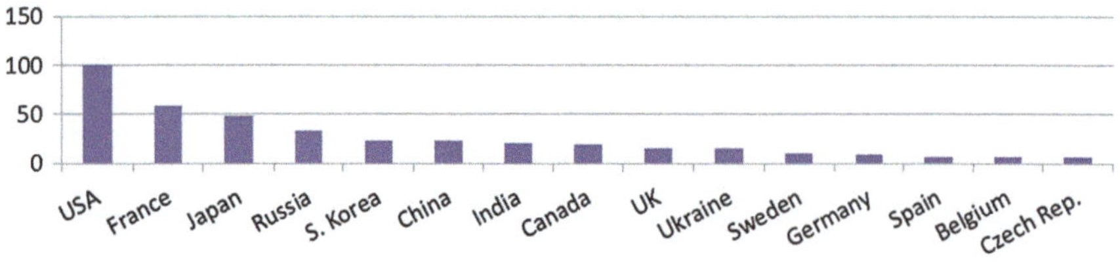

Fig. 18.1 Major Nuclear States: Total number of reactors [6]

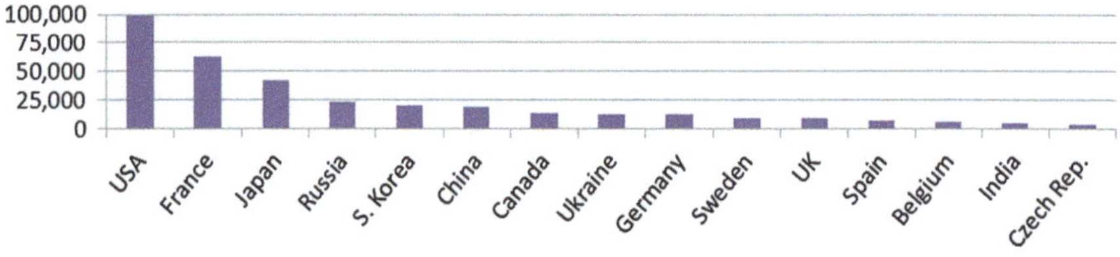

Fig. 18.2 Major Nuclear States: Total net electrical capacity (MW) [6]

18.3 Regulatory Oversight

Evaluations of the Fukushima accident have raised questions about regulatory oversight [1, 3, 7]. A study commissioned by the Japanese national legislature, for example, attributed a lax safety environment preceding the accident to a flawed safety culture and lack of regulator autonomy [3]. A 2013 meeting of national nuclear safety regulators from more than 50 countries articulated broader conclusions, recommending that nuclear safety regulators should utilize the International Atomic Energy Agency's (IAEA) peer review process as early as possible; and that peer reviews must include national action plans and follow-up missions [8, 9].

To clarify, engagement with the IAEA for its *Integrated Regulatory Review Service* (IRRS) currently serves as one of the more widely recognized forms of peer review to evaluate strengths and weaknesses of national regulatory entities, including organizational independence. IRRS assessments can be a useful tool, but must be recognized for their limitations to a nuclear safety regime. Reviews are done on a voluntary basis and follow-up assessments may not be completed. Reviewers also vary and, as with any subjective assessments, determinations are subject to interpretation.[2] Among the six countries of this study, all have undergone the IRRS process, except India, which will do so in 2015. France, Russia and the US – some

[2]Notably, Japan, underwent an IRRS in 2007 with findings that (1) its regulator was not independent, and (2) the organizational arrangements could "cause complexity". The IRRS also highlighted that the relationship management program of the Japanese regulatory body was well-structured and comprehensive, reflecting best practices [10].

Table 18.1 IAEA reviews: Integrated Regulatory Reviews Service (IRRS)

	China	France	India	S. Korea	Russia	US
Pre-Fukushima	1	2	0	0	1	1
Post-Fukushima	0	1	0	1	1	1
Planned			1	1		

Source: IAEA [11], as of November 2014

of the more traditional nuclear states – have been evaluated prior to and following the accident at Fukushima [11] (Table 18.1).

Functioning somewhat differently, from IRRS assessments the *Convention on Nuclear Safety* (CNS) requires contracting countries to agree to safety obligations and peer reporting review. This multilateral treaty, created in 1994 and now with 77 contracting parties, obliges member states to ensure that a regulatory framework is in place alongside a regulatory entity that has adequate authority, resources, and competence to oversee nuclear safety [12]. Member states must assure effective separation between safety oversight and industry promotion as well as other interests [7]. Since the Fukushima accident, there have been two review meetings and one extraordinary meeting in which contracting parties have reported on treaty obligations and discussed ways to improve nuclear safety worldwide. All six countries of this study are parties to the CNS and have participated in some manner. Like the IRRS process, CNS engagement serves a purpose, yet has limits on the extent of peer review.

18.3.1 Country Specifics

Engagement in IRRS and CNS reviews offers some indication of state-level commitments to nuclear safety, yet closer inspection of national regulatory bodies and related rules will provide greater depth on country specifics.

China *The Chinese nuclear regulator, National Nuclear Safety Administration (NNSA), has not been significantly restructured following the Fukushima accident. However, the accident underscored the importance of the agency. Additional financial and personnel resources have been made available. A mixed picture exists in terms of regulatory independence.*

The current NNSA in China was established in 2008, superseding its predecessor as the regulator of nuclear and radiation safety, and radiological environment management [13].[3] This agency acts as an administrative subordinate to the Ministry of Environmental Protection (MEP). In line with this, a 2010 IRRS review found that the NNSA "lacks some flexibility, within the policies and procedures of the

[3]The nuclear regulatory structure in China has undergone change in 1984, 1998, 2003, and 2008 [13].

government, to be competitive for recruiting experienced staff." However, specific to independence of the NNSA, the review did not find any issue "that could potentially put undue influence on its regulatory decision-making [13]." Other observers have, however, questioned the NNSA's autonomy. As a subordinate of the MEP, the NNSA is administratively positioned in the state apparatus beneath the state-owned NPPs which it oversees [14]. The NNSA also does not have its own R&D team to evaluate reactor designs for acquired technology (Ibid, citing personal communications). In 2011, the Chinese State Council Research Office recommended restructuring in such a way that the NNSA would be directly under the State Council Bureau, making it an independent regulatory body with greater authority [15].

Looking beyond the NNSA's organizational structure, China has been working in other ways to strengthen nuclear safety. Following the Fukushima accident, the pace of China's rapidly growing nuclear program was halted as a precaution to fortify its regulatory authority [7]. In line with this, licensing of new reactors was suspended in 2011 until the NNSA was improved, and until safety inspections were completed (Ibid). The Chinese government also consented to a budget increase for the NNSA and to a near doubling of its staff (Ibid). The 12th Five Year Plan recommended investment of nearly 80 billion RMB ($13 billion) by 2015 to enhance safety of NPPs, as robust growth occurs with different reactor technologies and designs, and safety standards ([16], citing 12th FYP).

Going further, little is broadly known about the appointment and removal processes of Chinese nuclear safety regulators. Currently, China does not have an atomic energy law, governing the use of the energy and associated activity. Such legislation has been under consideration since the 1980s and a draft is reportedly underway [16].

France *The French nuclear safety agency, Autorité de sûreté nucléaire (ASN), underwent notable change prior to the Fukushima accident. The shift is consistent with the lessons of independence tied to the accident.*

France established the ASN, an independent administrative authority on nuclear safety, with the Act on Transparency and Security (n° 2006 686, June 13, 2006) (TSN). The ASN is comprised of a board of five Commissioners which reports to Parliament. Commissioners are appointed for a non-renewable 6 year term and are not dismissible [17]. Following the establishment of the ASN, an IRRS was completed in 2006 with a subsequent review completed in 2009 at the request of the French government. In terms of agency independence, the 2006 review indicated: "In order to fully clarify and enhance its independent status, and put into place the new enforcement powers, ASN should as soon as practicable fully implement the requirements and powers given to it by the new TSN 2006 Act through elaboration and implementation of the necessary Decrees and Orders [18]." The subsequent review in 2009 found that the ASN had made significant progress in the implementation of requirements and powers granted to it in the 2006 Act. An IRRS in November 2014 may confirm this.

Changes in French nuclear safety rules appear to be limited post-Fukushima, in part since many safety improvements were introduced following a nuclear incident in France in 1999 [7].[4] The NPP operator is also partly state-owned, so formal regulation may not be required in the same way as with privately owned plants.

India *India's nuclear regulator, by many accounts, is not an autonomous body. Calls for restructuring were made prior to and after the accident at Fukushima. A proposal to restructure has been open at least since 2011.*

Established in 1983, the Indian Atomic Energy Regulatory Board (AERB) is a governmental board that reports to the Atomic Energy Commission where the utilization of nuclear energy is promoted. The AERB has been the subject of regulatory questions for quite some time. A Committee report in 1997 called for a revision to the Atomic Energy Act to improve regulatory efficacy [19]. In 2011, Parliament introduced a bill to create a new regulatory agency with continued questions about autonomy. In 2012, the Comptroller and Auditor General released a report, noting the lack of independence by the AERB, and findings that 60 % of regulatory inspections of operating NPPs were either delayed up to 153 days or not completed [20]. Smaller radiation facilities were also said to operate without rules or oversight (Ibid). In December 2013, a bi-partisan Public Accounts Committee Report described the AERB as "weak, under-resourced, and 'slow in adopting international benchmarks and good practices in the areas of nuclear and radiation and operation' (Ibid)." The report also noted that the AERB could not set or enforce rules for nuclear or radiation safety, and in many instances rules do not exist. The timeline for creating a new regulator remains to be seen a recent CNS report by India indicates that a bill to establish a new nuclear safety regulator is undergoing a final round of approval [21].

The current Chairman of the AERB, a former senior executive director of the nuclear power utility, noted in 2013 that AERB processes were evolving and challenged by the growth in new projects of diverse designs, a fast growing program, and operation in a competitive environment [22].

In July of 2014, rules relating to atomic energy were consolidated [23]. India's first IRRS is scheduled for 2015. The lack of separation between the AERB and the agency responsible for promoting the industry has been raised as a violation of the CNS [24]

Russia *Russia's nuclear safety regulator, Rostechnadzor, reports to the government and not to a specific ministry. A notable change occurred in its independence prior to the Fukushima accident. A Post-Fukushima review of the NPP fleet by the regulator and related team showed that the regulatory framework governing nuclear and radiological activity required revision for severe accident conditions and their prevention.*

The Federal Environmental, Industrial and Nuclear Supervision Service of Russia – *Rostechnadzor* – is an executive body that reports directly to the Government of the Russian Federation. Its current structure was revised in 2010, moving it from a formerly subordinate position, reporting to the Ministry of Natural Resources

[4] Vulnerabilities were found with the French Blayais plant in terms of flooding [7].

and the Environment. The autonomy of Rostechnadzor, according to a recent IRRS, is now "effectively independent of other federal executive authorities" with the removal of restrictions on its inspections [25]. Specific to appointments, a 2009 IRRS noted that the head of Rostechnadzor was appointed by the Prime Minister, and can be dismissed similarly. Decisions of Rostechnadzor could also be overturned by a court of law [26]. It is unclear if these organizational features were changed with the organizational revisions in 2010.

In terms of oversight activity, the 2013 IRRS indicated that the adequacy of human and financial resources reflected progress since 2009. However, more would be required to ensure the competency of the regulator [25]. Currently, the staff of Rostechnadzor relies on external entities for technical review and assessment of applications for approval. In line with aims to enhance the agency's capacity, a comprehensive management system is being implemented, and must advance in order to sufficiently support the now independent regulator (Ibid).

Following Fukushima, Rostechnadzor conducted a review of the Russian NPP fleet, finding that the regulatory framework which governs nuclear and radiological activity required revision for severe accident conditions and their prevention. In addition to its domestic focus, Russia has taken an active and leading role in improving the international regulatory regime on nuclear safety.

South Korea *The Nuclear Safety and Security Commission, South Korea's independent nuclear regulator, was established after the Fukushima accident. Reasons for its creation go well beyond the lessons of the accident in Japan.*

South Korea's Nuclear Safety and Security Commission (NSSC) was launched in late 2011 to oversee safety, security, and safeguards [27]. It is an independent entity which reports directly to the President of the Republic [7]. With its establishment, functions were separated from the promotion of industry.

The 2011 IRRS that was completed before the launch of the new NNSC noted issues in the commingling of safety regulation and industry promotion within one ministry. The report indicated a view that expected changes with the soon-to-be-created regulator were seen as a positive step with the new agency to have de jure and de facto independence [28].

As South Korea engages more prominently in the role of a developer and exporter of nuclear technology, domestic issues – including 215 cases of falsified quality records among nuclear technology suppliers and a safety failure of an NPP that went unreported for a month (Ibid) – highlight the continued need for a strong authority. A follow-up IRRS in December of 2014 should evaluate this.

USA *The U.S. Nuclear Regulatory Commission (NRC) is considered to be independent and has not undergone notable restructuring following the Fukushima accident. Recent changes have principally occurred in the make-up of its commissioners.*

Established in 1974, the U.S. NRC is a quasi-judicial body that is comprised of five commissioners and is widely deemed to be independent.[5] Commissioners are appointed by the President and confirmed by the Senate for 5-year terms with one individual being designated by the President to be the Chairman and official spokesperson of the Commission [30]. The status of the regulatory body has remained largely unchanged following the Fukushima accident. However, there have been political differences over how the Commission has been managed with personnel changes occurring in recent years. A 2010 IRRS indicated no concerns relating to the agency's independence. However, it did suggest that the NRC "assess whether the current regulations adequately provide for an independent verification of the safety assessment under the responsibility of the licensee before its use or submittal to the regulatory body and whether this verification is adequately confirmed by the NRC [31]." Specific to broader lessons from Fukushima and regulatory oversight, there has been a significant divergence of views within the organization over how venting requirements should be addressed with the NPP fleet in the US (*see* Sect. 18.4.3).

18.4 Filtering Technology in the Context of Safer Venting

18.4.1 Venting

The importance of venting during severe nuclear accident conditions ties to an ability to cool an NPP's core and containment system, as well as to control the pressure of the containment system. During emergency conditions, venting may preserve the integrity of an NPP's containment area (generally a reinforced steel or lead structure enclosing the reactor) by minimizing built-up hydrogen that can be produced during such an event. Venting also allows for the reduction of steam within the containment system, so that water can be pumped in to cool the fuel rods.

Specific to the Fukushima accident, delays in venting were associated with uncertainty about reactor conditions and the decision-making process, followed by subsequent difficulties in physically carrying out the decision to open vents [32]. The Fukushima Daiichi Mark I and Mark II BWR design was particularly vulnerable, as noted earlier, due to its relatively small containments, and dependence on suppression pools that do not mitigate hydrogen [33]. If containment venting had occurred during earlier stages of the accident, some of the more detrimental

[5]While the NRC is based within the executive branch of the President, the President does not typically intervene in decisions but does appoint commissioners. The NRC's budget is funded principally by fees charged to regulated entities ([24] citing Malsch, 2011 and Borchardt 2009). Congress wields some influence as it approves nominees, can change the law governing the Commission, and appropriates funds [24]. The NRC has an internal advisory committee as an independent body charged with double-checking licensing decisions [29].

consequences may have been averted. The explosions and subsequent damage to the NPP released significant amounts of radiation into the environment.[6]

18.4.2 Filtering Technology

In the period since the accident, considerable attention has centered on filtered containment venting systems (FCVS). Such filtering technology can substantially improve mitigation efforts by reducing the level of radioactive release. In a scenario of excess pressure, particularly requiring rapid containment, venting with filtration can help to protect a population that is unable to quickly evacuate, as well as protect the surrounding environment. Compared to options like hardened vents and severe accident confinement strategies, FCVS is deemed to provide more regulatory certainty and allows for timely implementation [35].

Such technology is not new. Subsequent to the Three Mile Island (TMI) accident in 1979 and that at Chernobyl in 1986, FCVS was installed in countries such as Finland, France, Germany, Switzerland and Sweden [36]. In the late 1990s through to 2011, additional countries like Bulgaria, Canada and the Netherlands evaluated and in some cases back-fitted the technology (Ibid). FCVS is now to be installed in the Japanese NPP fleet that remains operational [35] and is required in many countries [37]. Given the nature of the international nuclear playing field, there is, however, no international standard which requires that FCVS be used in BWRs. The majority of countries, which utilize Mark I and Mark II BWRs, have revised or are in the process of modifying the design to include FCVS. Some states are going even further to require FCVS in NPPs, other than Mark I and Mark II BWRs [37].[7]

18.4.3 Country Specifics on FCVS and Related Technology

Specific to the six countries of this study, there is wide variation in terms of requirements and implementation of FCVS and related technology.

China China, which currently has a fleet of 23 NPPs and is constructing 26 more (principally Pressurized Water Reactors (PWRs)) may consider the implementation of FCVS for some reactors ([7]; personal communications with NNSA 2014).

[6]Estimates indicate that releases of Iodine 131 and Cesium 137 in the Fukushima accident equaled ~10 % and ~20 %, respectively, of that for the Chernobyl accident, based in averages of published numbers [34].

[7]With FCVS technology, major types include: Sand-bed filter with dry metallic pre-filter (EDF/IRSN); Multi-Venturi scrubber (Toshiba/Westinghouse); Dry filter method (Toshiba/Westinghouse); High-speed sliding Venturi (Siemens-Areva); and second generation Wet Scrubber (Sulzer/IMI) [36]. Variations exist in efficiency and lay-out.

France All 58 operating NPPs in France are equipped with the sand-bed form of FCVS [37]. Following the publication of the WASH 1400 report [38] and the accident at TMI, France adopted the use of FCVS in all its reactors [37]. After the Fukushima accident, the ASN directed EDF to submit a detailed study of potential improvements to its venting-filtration system, accounting for the potential necessity of filtering multiple reactors simultaneously [7]. Enhanced FCVS are now being evaluated to satisfy such requirements in the 58 Generation II PWRs. No decision has yet been made to implement such upgrades [37]. The single Generation III EPR that is under construction has a built-in Containment Heat Removal System which relies on a spray system that is dedicated to severe accidents. AN FCVS option has been studied and the utility has proposed to use mobile equipment for containment spray in the EPR.

India India currently does not require FCVS for its 21 NPPs or for the 6 under construction [35]. The fleet includes BWRs (of the Mark 2 design), PWRs, PHWRs, and an FBR under construction. Recommendations to install FCVS were put forward by an appointed committee (Personal communications with [23]). The utility has submitted a design basis report for modification (Ibid).

Russia Currently, there is no national regulatory requirement for FCVS in Russia, and the technology is not installed on any of the 33 operating NPPs, which include an FBR, VVERs, RBMKs and graphite moderated BWRs [37]. Following the Fukushima accident, FCVS is being considered for some VVERs (Ibid).

South Korea South Korea completed the installation of FCVS (High Speed Sliding Pressure Venturi type) in one PHWR (Wolsong 1) in 2012. The rest of the 23 NPPs, including PWRs and PHWRs, are in the process of installing the technology before 2019 through an open tendering process (Ibid).

USA The US does not currently require FCVS in any of its 100 operating NPPs. Its fleet is comprised of 65 PWRs and 35 BWRs, including the Mark I and Mark II design. None of the NPPs have FCVS installed [37]. Subsequent to the Fukushima accident, the NRC ordered 31 reactors with the design similar to that in Fukushima to retrofit reliable hardened containment vents over two stages [30]. Phase 1 provides operators a period up to June 2018 to install a hardened vent (severe accident capable) on the wet-well containment system. Phase 2 provides operators a period up to June 2019 to install a hardened vent on the dry-well containment venting system or to develop a "reliable containment venting strategy that makes it unlikely that a licensee would need to vent from the containment" [35].

As noted earlier, differences exist within the NRC on how FCVS should be handled. In 2012, the NRC staff put forward technical recommendations, advising that operators of Mark I and Mark II BWRs be ordered to modify pressure suppression containment systems with FCVS (external filter) connected to wet-well and dry-well components [35, 39]. A majority of Commissioners voted in March 2013 against issuing an immediate order and instead requested that the staff produce a technical evaluation to support the rule-making through the regular process as opposed to as an emergency issue [7, 40]. The Staff was directed to study the use of a filter added to a vent, and a performance-based approach using incumbent systems to attain a similar reduction in radiation release during emergency conditions (Ibid). Determination of the way forward remains an open question.

18.5 Conclusion

Reviewing the status of nuclear safety regulatory entities post-Fukushima in six, major nuclear states, one finds varying degrees of autonomy that are evident. Here, important questions persist. Will staffing and resource changes sufficiently equip the Chinese and Russian regulators to manage the rapid scale-up of their nuclear programs? How will South Korea's new regulator be assessed in the December 2014 IRRS, given its recent issues with fraud and lax reporting, alongside national growth as an exporter of NPPs? How will the U.S. NRC manage its organizational differences over its rule-making? Finally, will India's AERB be restructured with greater autonomy and be adequately equipped to allow for appropriate oversight as its nuclear program advances?

Specific to FCVS, the Fukushima accident highlighted that such technology could be pivotal in minimizing explosions and radiation releases. The technology is not new and has been adopted in many countries. Among the group of six countries in this study, France uses the technology, South Korea is adopting it, and the other four countries have yet to settle this. Considered in the contest of regulatory and voluntary adoption, a number of questions remain. First, how do reactor designs other than Mark I and Mark II BWRs (which are not as challenged in their design) optimally meet aims consistent with the goals of FCVS? Can the benefits of filtering be attained with means other than FCVS? How do costs of multiple FCVS designs compare?

In thinking more broadly across the nuclear playing field, areas to watch include India's program which is scaling up, has BWRs without FCVS, and is overseen by a regulator with questions of efficacy. Other programs naturally also must remain vigilant. Can regulatory learning and capacity building keep pace with NPP fleet growth? Moreover, will the time to decide on FCVS and/or to strengthen the regulator leave the safety regime at risk?

The door remains open on all of these questions.

References

1. Independent Investigation Commission on the Fukushima Nuclear Accident (IIC), Bricke M (eds) (2014) The Fukushima Daiichi nuclear power station disaster. Earthscan, London
2. Institute of Nuclear Power Operations (INPO) (2012) Lessons learned from the nuclear accident at the Fukushima Daiichi nuclear power station, INPO 11-005, Addendum, August 2012

3. National Diet of Japan (Diet) (2012) The official report of the Fukushima nuclear accident independent investigation commission (Diet Report). https://www.nirs.org/fukushima/naiic_report.pdf
4. Nuclear Energy Agency (NEA) Organization for Economic Cooperation and Development (OECD) and Committee on Safety of Nuclear Installations (CSNI) (2013). The Fukushima Daiichi nuclear power plant accident, report
5. National Research Council (NRC) (2014) Lessons learned from the Fukushima nuclear accident for improving safety of U.S. nuclear plants. National Academies Press, Washington, DC
6. International Atomic Energy Agency (IAEA) (2014) PRIS database
7. Government Accountability Office (GAO), United States (2014) Nuclear safety, report to the chairman, subcommittee on transportation and infrastructure, committee on environment and public works, U.S. Senate, GAO 14 109
8. International Atomic Energy Agency (IAEA) (2013) International Conference on Effective Nuclear Regulatory Systems, http://www-pub.iaea.org/iaeameetings/41986/international-conference-on-effective-nuclear-regulatory-systems
9. International Atomic Energy Agency (IAEA) (2013) Conference on regulatory effectiveness. http://www.iaea.org/newscenter/news/conference-regulatory-effectiveness-concludes
10. International Atomic Energy Agency (IAEA) (2007) Integrated Regulatory Review Service (IRRS) Mission to Japan, Tokyo, Japan, 25–30 June 2007
11. International Atomic Energy Agency (IAEA) (2014) IRRS Worldwide, http://gnssn.iaea.org/regnet/irrs/Pages/IRRS-Worldwide.aspx
12. International Atomic Energy Agency (IAEA) (1994) Convention on nuclear safety, vol 16, Legal series. IAEA, Vienna
13. International Atomic Energy Agency (IAEA) (2010) Integrated Regulatory Review Service (IRRS) Mission to People's Republic of China, Beijing, People's Republic of China, 18–30 July 2010
14. Zhou Y, Rengifo C, Chen P, Hinze J (2011) Is China ready for its nuclear expansion? Energy Pol 39:771–781
15. World Nuclear Association, Government Structure and Ownership. http://www.world-nuclear.org/info/Country-Profiles/Countries-A-F/Appendices/Nuclear-Power-in-China-Appendix-1--Government-Structure-and-Ownership/. September 2014
16. Nakano J (2013) The United States and China: making nuclear energy safer, Brookings paper, Center for Strategic and International Studies, July 2013
17. Lacoste A (2013) Nuclear safety and radiation protection regulation in France, Presentation, Taiwan, 29 Jan 2013
18. International Atomic Energy Agency (IAEA) (2009) Integrated Regulatory Review Service (IRRS) Follow-up Mission to France, Paris, France, 29 Mar to 3 Apr 2009
19. Happymon J (2014) Regulating India's nuclear estate, The Hindu. 29 August 2014
20. Doherty B (2013) Harsh criticism for India's nuclear safety regime. http://www.smh.com.au/world/harsh-criticism-for-indias-nuclear-safety-regime-20131220-hv6lz.html
21. Convention on Nuclear Safety (CNS) Reports by respective governments for France (2013/2014), India (2013), Russia (2014), South Korea (2013), and the US (2013)
22. Bajaj SS (2013) Regulatory practices for NPPs in India. Sadhana 38(5):1027–1050
23. Atomic Energy Regulatory Board (AERB) (2014) Government of India, policies governing regulation of nuclear radiation safety, report, July 2014
24. Bacon-Dussault M (Undated) Independence of nuclear regulators in the aftermath of the Fukushima Daiichi nuclear accident. Canadian Nuclear Safety Commission. Ottawa
25. International Atomic Energy Agency (IAEA) (2013) Integrated Regulatory Review Service (IRRS) Follow-up Mission to the Russian Federation, Moscow, Russian Federation, 11–19 Nov 2013
26. International Atomic Energy Agency (IAEA) (2009) Integrated Regulatory Review Service (IRRS) Mission to the Russian Federation, Moscow, Russian Federation, 16–27 Nov 2009

27. Nuclear Law Bulletin (NLB). 88: 2011/2, 87
28. International Atomic Energy Agency (IAEA) (2011) Integrated Regulatory Review Service (IRRS) Mission to the Republic of Korea, Daejon, Republic of Korea, 10–22 July 2011
29. Bredimus A, Nuttal W (undated) A comparison of international regulatory organizations and licensing procedures for new nuclear power plants, Judge Business School. University of Cambridge, Cambridge
30. Nuclear Regulatory Commission (NRC) (2015) The Commission, http://www.nrc.gov/about-nrc/organization/commfuncdesc.html
31. International Atomic Energy Agency (IAEA) (2010) Integrated Regulatory Review Service (IRRS) Mission to the United States, Washington, DC, 17–29 Oct 2010
32. Nuclear Energy Institute (NEI) (2012) Vents and filtering strategies come to forefront in Fukushima response. http://www.nei.org/News-Media/News/News-Archives/Vents-and-Filtering-Strategies-Come-to-Forefront-i
33. Leyse M (2012) Post-Fukushima hardened vents with high-capacity filters for BWR Mark I's and Mark II's, report, July 2012
34. United Nations Scientific Committee on Effects of Atomic Radiation (UNSCEAR) (2013) Report to the general assembly, levels and effects of radiation exposure due to the nuclear accident after the 2011. Great East Japan earthquake and tsunami, vols I and II
35. Nuclear Regulatory Commission (NRC) (2012) Consideration of Additional Requirements for Containment Venting Systems for BWRs with Mark I and Mark II Containments, SECY 2012 0157
36. Hillrichs C, Vignon D, Felten P (2012) Review of European filtered containment venting systems, Report, ENC
37. Nuclear Energy Agency (NEA) (2014) Committee on the safety of nuclear installations, OECD/NEA/CSNI status report on filtered containment venting, 2 July 2014, NEA/CSNI/R (2014)7
38. Nuclear Regulatory Commission (NRC) (1975) Reactor safety study, WASH 1400 (NUREG 75/014), Oct 1975
39. Gunter P et al (2014) Request for an inquiry into commission actions rejecting the NRC staff recommendation (option 3) to order all Mark I and Mark II BWR operators install engineered external radiation filters on hardened
40. Nuclear Regulatory Commission (NRC) (2013) EA 2013 109

Chapter 19
The Value of Standards for Detection of Radioactive Materials

Leticia Pibida and Anne Sallaska

Abstract Document and physical standards play a vital role in security applications. Standards allow setting common measurement units, physical quantities and performance requirements for different types of radiation detection equipment and devices. Document standards provide a common ground for allowing consensus between user requirements, manufacturing process and testing capabilities. Physical standards address the reproducibility and traceability of measurements to national and/or international standards. Different standards have been developed for different types of instruments used in detection of illicit trafficking of radioactive materials and homeland security applications. These document standards are being used as part of a laboratory accreditation program that could be used for the development of an instrument certification program.

19.1 Introduction

Radiation detection instruments are used in many different applications for detection and identification of radioactive materials that may pose a nuclear or radiological threat. Users of radiation detection instruments may have different requirements depending on the environment in which these instruments could be used as well as the radiation field level that is expected to detect. There are a wide variety of commercially available instruments from which to choose. These types of instrument range from small pocket size (e.g., personal radiation detectors (PRDs)), to hand-held instruments (e.g., radionuclide identification detectors, gamma-ray survey meters, and neutron detectors), body-worn instruments (e.g., backpack-type detectors) as well as large installed or mobile systems (e.g., portal monitors and vehicle mounted systems). The selection of an instrument will depend on the type of deployment. The types of deployments may vary widely, from law-enforcement officers in a major event (e.g., Olympic Games, conventions) to customs officers, military personnel and coast guards. In order to make an informative decision about

L. Pibida (✉) • A. Sallaska
National Institute of Standards and Technology, Gaithersburg, 27899-8462, MD, USA
e-mail: leticia.pibida@nist.gov

S. Apikyan, D. Diamond (eds.), *Nuclear Threats and Security Challenges*,
NATO Science for Peace and Security Series B: Physics and Biophysics,
DOI 10.1007/978-94-017-9894-5_19

the instrument capabilities it is necessary to be able to compare the performance of the different types of available instruments. Testing of radiation detection instruments against standards is an initial step that allows the assessment of the instrument performance. Standards can be divided in two classes, document and physical standards. The document standards normally provide a set of requirements and associated test methods that are based on user requirements and the capabilities of current technology. The physical standards are objects or tools such as radioactive sources that are used as part of the instrument testing procedure. Standards are needed because they provide specifications and procedures to ensure reliability of materials, products, methods and/or services used by people.

Standards are important because they provide safety and reliability in a product raising user confidence in the performance of instruments. They support government policies and legislation, allow for interoperability of devices and provide a foundation for the development of new features and options required by consumers. Businesses and manufacturing also benefit from standards as they allow them to:

- Develop new technologies
- Enhance existing practices
- Open up market access
- Encourage innovation
- Increase awareness of technical developments

In this paper we briefly describe the different types of standards, the issues to be considered when developing a standard and the implications of their use for the assessment of the different types of radiation detection instruments used for the detection of nuclear and radiological threats.

19.2 Types of Standards

As mentioned in the introduction, there are mainly two types of standards; document and physical or measurement standards. The physical or measurement standards are objects (e.g., standard reference materials, radioactive sources), structures or methodologies used for determining the performance of a specific type of instrument or measurement method. Document standards provide guidance, requirements and test methods to evaluate the instrument performance against a given set of requirements.

Several document standards were developed for instruments used in the detection and identification of nuclear and radiological threats by the American National Standard Institute/Institute of Electrical and Electronics Engineers (ANSI/IEEE) and the International Electrotechnical Commission (IEC) organizations. A list of the currently published ANSI/IEEE and IEC standards for instruments used for detection and identification of illicit trafficking of radioactive materials and homeland security applications are listed in Table 19.1. These standards were developed for detection of nuclear and radioactive materials except for the ANSI

Table 19.1 List of published ANSI/IEEE and IEC standards for instruments used for detection of illicit trafficking of radioactive materials and homeland security applications

ANSI/IEEE standards	IEC standards
ANSI/IEEE N42.32	IEC 62401
Performance criteria for alarming personal radiation detectors for homeland security	Radiation protection instrumentation – alarming personal radiation devices (PRD) for detection of illicit trafficking of radioactive material
ANSI/IEEE N42.33	IEC 62533
Portable radiation detection instrumentation for homeland security	Radiation protection instrumentation – highly sensitive hand-held instruments for photon detection of radioactive material
ANSI/IEEE N42.34	IEC 62327
Performance criteria for hand-held instruments for the detection and identification for radionuclides	Radiation protection instrumentation – hand-held instruments for the detection and identification of radionuclides and for the indication of ambient dose equivalent rate from photon radiation
ANSI/IEEE N42.35	IEC 62244
Evaluation and performance of radiation detection Portal monitors for use in homeland security	Radiation protection instrumentation – installed radiation monitors for the detection of radioactive and special nuclear materials at national borders
ANSI/IEEE N42.38	IEC 62484
Performance criteria for spectroscopy-based portal monitors used for homeland security	Radiation protection instrumentation – spectroscopy-based portal monitors used for the detection and identification of illicit trafficking of radioactive material
ANSI/IEEE N42.42	IEC 62755
American national standard data format for radiation detectors used for homeland security	Radiation protection instrumentation – data format for radiation instruments used in the detection of illicit trafficking of radioactive materials
ANSI/IEEE N42.43	IEC 62534
Performance criteria for mobile and transportable radiation monitors used for homeland security	Radiation protection instrumentation – highly sensitive hand-held instruments for neutron detection of radioactive material
ANSI/IEEE N42.48	IEC 62618
Performance requirements for Spectroscopic Personal Radiation Detectors (SPRDs) for homeland security	Radiation protection instrumentation – Spectroscopy-based alarming Personal Radiation Detectors (SPRD) for the detection of illicit trafficking of radioactive material
ANSI/IEEE N42.53	IEC 62694
Performance criteria for backpack based radiation detection systems used for homeland security	Radiation protection instrumentation – Backpack-type radiation detector (BRD) for detection of illicit trafficking of radioactive material
ANSI/IEEE N42.49A	
American national standard for performance criteria for alarming electronic Personal Emergency Radiation Detectors (PERDs) for exposure control	

N42.49A standard which was specifically developed to be used in response to a radiological incident. These standards do not address instrument requirements or measurement needs when recovering from a radiological or nuclear event. The ANSI/IEEE N42.42 standard and the IEC 62755 are used by all the other instrument standards as they define the data format requirements for radiation detection instruments.

The National Institute of Standards and Technology (NIST) developed different types of radioactive sources to test radiation detection systems addressed in several of these standard documents [1, 2]. The sources were designed with different objectives in mind. Some of these sources are currently commercially available with the activity values (in units of Bq) required in the standards traceable to the national standard held at NIST [1].

19.3 Document Standards Development Organizations

Document standards can be developed by different groups and organizations. Different development groups may have different focuses and needs depending on the users they represent. Some groups concentrate more on the performance requirements of a particular type of instrument, while others provide guidance for testing, use or calibration of instruments. Standards can also be developed to define specific design requirements for instruments. These groups and organizations include professional societies, trade associations, testing and certification organizations as well as national or international standards development organizations (e.g., IEC, ANSI, IEEE). The following list provides examples of such standards organizations that developed document standards for radiation detection instruments (not all inclusive):

- ANSI: American National Standard Institute
- ASTM: American Standards for Testing and Materials
- DOD: Department of Defense
- HPS: Health Physics Society
- IAEA: International Atomic Energy Agency
- IEC: International Electrotechnical Commission
- IEEE: Institute for Electrical and Electromechanical Engineers
- ISO: International Standards Organization

The standards organizations such as ANSI or IEC have different committees under which standards are developed. The standards listed in Table 19.1 were developed under the ANSI Standards N42 Committee on Radiation Instrumentation [3] and the IEC Technical Committee 45 Subcommittee B Working Group B15 (TC45B WGB15) [4].

19.4 Issues Considered During the Standards Development

The main principles behind the development of document standards are:

- Transparency – need to ensure that they are accessible to all interested parties
- Openness – have a broad participation from different sectors that are interested in a specific product/instrument
- Impartiality – ensure that the standard does not favor a particular product or manufacturer
- Effectiveness and Relevance – standards are normally developed as part of a response to a need
- Consensus – require agreement between involved parties as to the content in the standard and be developed in a collaborative manner
- Technically Based – require rigor in requirements and test methods

The standards listed in Table 19.1 cover a large number of requirements which are generally grouped in separate categories including general, radiological, environmental, mechanical and electromagnetic requirements. The general requirements include basic instrument features such as display, weight, size, alarm types, data format and user interface. The radiological requirements include tests such as false alarm rates, gamma and neutron response, time to alarm, accuracy, over-load, background effects, neutron detection in the presence of photons, and radionuclide identification (i.e., single radionuclide, shielding, simultaneous radionuclides). The environment requirements include temperature, humidity, cold and hot temperature start up, moisture and dust tests. The mechanical requirements include vibration, shock, drop and impact tests. The electromagnetic requirements include radio frequency, magnetic fields, electrostatic discharge, surges and oscillatory waves, radiated emissions, conducted emissions and conducted disturbances. The types of requirements addressed by these standards depend on the type of instrument and its intended use (e.g., portable or installed equipment). For each requirement the standard defines an acceptance criterion.

A main issue to consider in the development of a standard is the level of the instrument performance required by the users. The user requirements need to drive the design of the requirements and test methods in the standards. Normally users require having an instrument that displays a high probability of detection or identification of a radiological or nuclear threat with a high confidence level. This translates into a given number of trials and failures for a particular test. The larger the required probability the larger will be the number of trials and the higher the cost of testing. Therefore, there is a need to establish a compromise between the user expectations and the cost of performing the test. To illustrate this issue let us assume that the standard is trying to evaluate the instrument alarm response function under the following conditions: a radioactive source is passing by an instrument at a speed of 1.2 m/s and the distance traveled by the source is 5 m.When the source

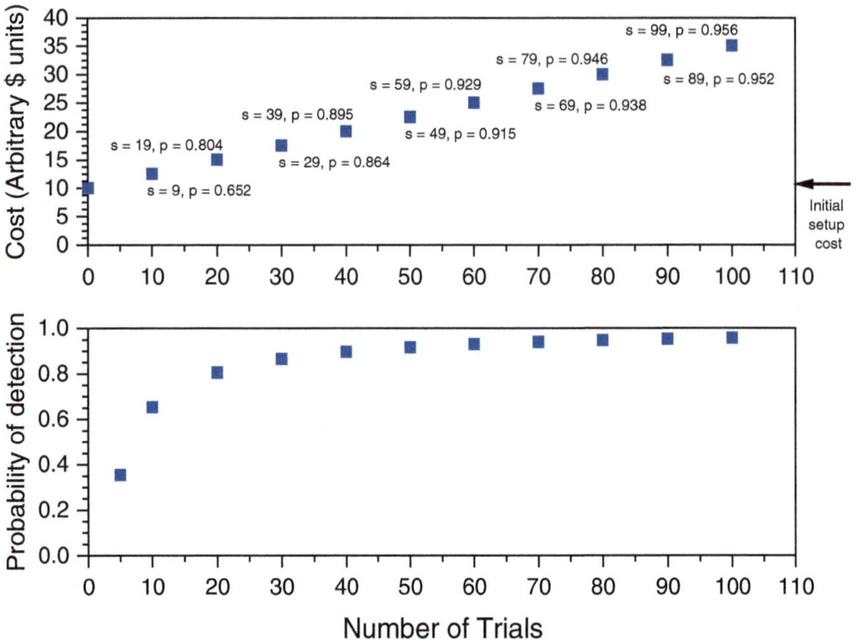

Fig. 19.1 *Top* – cost of testing as a function of the number of trials, where "s" is the number of successes and "p" is the probability of success. *Bottom* – probability of detection as a function of the number of trials assuming that the acceptance criterion allows for 1 failure for a 95 % lower confidence bound for a binomial distribution

passes in front of the instrument the person conducting the test will need to verify and record if the instrument generates an alarm. A single trial consists of a single pass of the source by the instrument. Assuming that there is a waiting time between trials of 10 s then the total time to conduct a single trial is about 14 s (4 s for the detector source to pass by the detector plus 10 s of waiting time). Let us assume that the cost of testing increases linearly with time and that the time it takes to setup the test is 10 min. Figure 19.1 shows the cost of testing as a function of the number of trials (top) and the increase in the probability of detection (or success) as a function of the total number of trails (bottom). In this particular case, the probability of detection is calculated based on an acceptance criterion that allows for one failure with a 95 % lower confidence bound for a binomial distribution. From this figure it can be observed that the increase in testing cost for running 40 trials is similar to that of the initial test setup. As the number of trials increase, the cost of testing increases linearly with the number of trials while the probability of detection seems to plateau. For most radiological tests described in the standards, the instrument acceptance requirement is based on ten trials and nine successes; this means that the probability of success is equal to approximately 0.65. For example, if tests will be performed for 30 trials while allowing for one failure then the probability of success will be approximately equal to 0.86. This is an increase of approximately 20 % in the probability of detections compared to the current requirements, while the cost of performing 30 trials will be lower than initial test setup cost. This example illustrates that for every test, an optimal number of trials can be chosen that balances the gain in knowledge about the instrument performance versus the cost of testing.

The environmental, mechanical and electromagnetic requirements depend on the intended conditions of use. The standards include broad ranges for these influence quantities to account for most possible conditions encountered by the different types of users of the instruments. The acceptance criterion for these types of tests allows the instrument radiation field readings to drift by a fixed number (e.g., ±15 %, ±30 %). For instruments with radionuclide identification capabilities it requires that the number of correct identifications remains unchanged or is better compared to the response obtained at a reference testing point. This acceptance criterion does not account for the uncertainty of the instrument radiation field readings and possible variations in the spread of these readings when exposed to the different influence quantities. New acceptance criteria are being presently discussed during the revision of some of the standards to account for these effects.

As shown in Table 19.1, for a given type of instrumentation there is usually an ANSI/IEEE standard for each IEC standard listed. For these cases, both the ANSI/IEEE and IEC standards have similar requirements for most tests. The main differences are in the required instrument performance (i.e., number of trials for a given test) and the physical quantities and units used to define the radiation field. In addition, the different ANSI/IEEE standards use different quantities to define the radiation field. These quantities include:

- Exposure rate, (ANSI/IEEE) – measured in units of μR/h
- Ambient dose equivalent rate, (IEC) – measured in units of μSv/h
- Activity – measured in units of Bq
- Emission rate – in units of s^{-1}
- Fluence rate – in units of $cm^{-2}{\cdot}s^{-1}$
- Mass and geometry – main used for special nuclear material (SNM) sources

For a given type of instrument, the main difference between the ANSI/IEEE and IEC standards is the use of exposure rate (in units of μR/h) and ambient dose equivalent rate (in units of μSv/h). As a result of this different use of units the actual value of the required testing fields are slightly different due to the required use of conversion coefficients to convert from one quantity to the other [5]. For example, for the personal radiation detectors (PRDs) the time to alarm test in the ANSI/IEEE standard requires testing the instruments using ^{241}Am, ^{137}Cs and ^{60}Co producing a radiation field of 50 μR/h at the reference point of the PRD while the IEC standard requires a field of 0.5 μSv/h. This corresponds to a difference of approximately 52 %, 5 % and 2 % for ^{241}Am, ^{137}Cs and ^{60}Co respectively. There are advantages and disadvantages when using different quantities to define the radiation field required for testing the radiation detection instruments, as summarized in Table 19.2.

Table 19.2 Advantages and disadvantages of using different quantities to determine the radiation field used for testing

Quantity	Advantages	Disadvantages
Activity	Fixed quantity and testing distance, can purchase calibrated sources with known uncertainty	Need to purchase sources when they decay
Exposure rate	Can use different source activities and testing distance	Requires a calibrated instrument with a known energy response to determine the exposure rate. Large uncertainty for 5 μR/h and 50 μR/h fields
Ambient dose equivalent rate	Can use different source activities and testing distance	Same as the exposure rate, may require applying conversion coefficients depending on the quantity displayed by the instrument
Emission rate	Same as for activity, good for specification of SNM and Depleted Uranium (DU) sources	Requires a calibrated high resolution detector (e.g., HPGe) for accurate measurements
Fluence rate	Can use different source activities and testing distance	Same as for emission rate
Mass/Geometry	Mainly used to SNM sources and DU. Fixed quantity, easy to use	Specific SNM mass quantities and geometries are not easy to find. Cannot change testing distance

19.5 Use of Standards

Standards are used by different organization for testing of radiation detection instruments. The most recent testing efforts were carried out by the Illicit Trafficking Radiation Assessment Program (ITRAP+10). This effort was started by the European Union to evaluate the performance of commercially available European manufactured radiation detection instruments against nine of the ANSI/IEEE and IEC standards listed in Table 19.1. The European Commission Joint Research Center (EC-JRC) invited the U.S. government and the International Atomic Energy Agency (IAEA) to participate in the ITRAP+10 to expand the testing to other manufacturers. The Department of Homeland Security (DHS) Domestic Nuclear Detection Office (DNDO) implemented the U.S. participation of this effort [6].

These standards are also used to test radiation detection instruments under the DHS/DNDO Graduated Rad/Nuc Detector Evaluation and Reporting (GRaDER®) program [7]. This program makes use of laboratories accredited by the National Voluntary Laboratory Accreditation Program (NVLAP) [8] to test radiation detection instruments against the ANSI/IEEE standards listed in Table 19.1 (with the exception of the ANSI/IEEE N42.49A standard).

Manufacturers use these standards in the development of new instruments as well as in the improvement of existing technology. Instrument users may utilize these standards as part of their procurement requirements and their instrument acceptance process.

19.6 Summary

Standards are a useful tool to compare instrument performance in a controlled environment type testing. Standards are living documents, as they evolve with new requirements and advances in the technologies. Several different variables need to be considered during the standard development including testing cost, instruments capabilities and possible deployment applications. Multiple organizations develop standards for systems or methods that are used worldwide. International standards play a major role in the development of standards that are common to many countries, and these countries may modify some of the international set of requirements based on specific country needs. Several ANSI/IEEE and IEC standards were developed for a variety of radiation detection instruments used in homeland security and detection of illicit trafficking of radiological and nuclear threats.

References

1. Lucas L, Pibida L, Unterweger M, Karam L (2005) Gamma-ray emitting test sources for portal monitors used for homeland security. Radiat Protein Dosim 113(1):108–111
2. Lucas L, Pibida L (2009) New spherical gamma-ray and neutron emitting sources for testing of radiation detection instruments. J Res Nat Inst Stand Technol 114(6), November–December 2009
3. ANSI/IEEE standards. http://www.ansi.org/
4. IEC standards. http://www.iec.ch/
5. ISO 4037-3, X and gamma reference radiation for calibrating dosemeters and doserate meters and for determining their response as a function of photon energy – Part 3: calibration of area and personal dosemeters and the measurement of their response as a function of energy and angle of incidence
6. ITRAP+10. http://www.dhs.gov/illicit-trafficking-radiological-assessment-program-10-itrap10
7. GRaDER® program. http://www.dhs.gov/guidance-grader-program
8. NVLAP. http://www.nist.gov/nvlap/

Chapter 20
Novel Nuclear Measurements Technologies for Safety and Security

Massimo Morichi, Roger Abou-Khalil, Philippe Dubart, and William Russ

Abstract A series of novel nuclear measurements technologies such as hybrid gamma spectroscopy and real time 3D-contamination modelling have been developed recently at AREVA towards Safety and Security applications as result of direct experiences and related to Fukushima site remediation project.

The novel hybrid spectroscopic systems, combining spectra from different type of detectors (NaI, HpGe, CdZnTe, CsI(Tl)..) (High and low-resolution detectors) are evaluated for security screening applications. Repeated measurements with a variety of sources and acquisition times are analysed, with the results showing that hybridization can enhance peak detectability relative to the individual constituent detectors, improve detection probability and reduce the false alarm rate. Fukushima accident imposed a stretch to nuclear measurement operational approach requiring in such emergency situation: fast deployment and intervention, quick analysis and fast scenario definition. AREVA, as return of experience from his activities carried out at Fukushima has developed a novel multi-sensor solution as part of his D&D research, approach and method, a system with real-time 3D photorealistic spatial radiation distribution cartography of contaminated premises. The system may be handheld, mounted on a mobile device (robot, drone, e.g.). In this paper, some details of the current development based on a SLAM technology (Simultaneous Localization And Mapping) and integrated sensors and detectors allowing simultaneous topographic and radiological data are showed. System allows dose rate and gamma spectrum distribution in a 3D activity mapping.

M. Morichi (✉) • R. Abou-Khalil • P. Dubart
AREVA Group, Courbevoie, France
e-mail: massimo.morichi@areva.com

W. Russ
CANBERRA an AREVA Company, Meriden, CT, USA

S. Apikyan, D. Diamond (eds.), *Nuclear Threats and Security Challenges*,
NATO Science for Peace and Security Series B: Physics and Biophysics,
DOI 10.1007/978-94-017-9894-5_20

20.1 Introduction

Radiometric systems used to screen and prevent illicit trafficking of radiological materials are designed with the goal of being very sensitive and selective. While high sensitivity of such non proliferation technologies enhances the probability of detection, selectivity is also needed to accurately corroborate or refute manifests of legal shipments and to better avoid false alarms caused by innocuous sources. Selectivity is best achieved using high resolution spectroscopy detectors, but such detectors are limited in size and by cost.

The efficiency of a spectroscopic measurement system can be improved by adding detectors to the system and combining all the signals. Historically, this has been done using detectors of the same type, all with the same peak response function. Summing measured spectra with matched energy calibrations allows analysis of the final summation spectrum using the common peak response function. The benefit is improved sensitivity at the cost of the additional detectors and additional calculations to match energy calibrations prior to spectral summing.

The traditional approach of combining systems of like detectors does not support combining different types of detectors with very different peak response functions. Standard summing of disparate detectors would require subsequent analysis of the summed spectrum using a complex multi-modal peak response function dependent on the relative efficiency ratio as a function of energy. A novel approach developed at AREVA has been proposed for combining spectroscopy detectors of any type, producing a hybrid spectrum [1] that can be analysed with a simple peak response function, accommodating existing analysis methods. This hybrid detection system would combine detectors of different types into a single spectroscopy system.

A hybrid detection system might combine a higher efficiency, lower resolution detector with a lower efficiency, higher resolution detector, achieving both higher efficiency and better energy resolution in a single system. Potential applications envisioned are related to low level waste characterization, internal contamination measurements, as well as non-proliferation and safeguards and more generally all applications that require simultaneously higher detection efficiency and higher resolution.

To evaluate the efficacy of such a hybrid system designed for non-proliferation applications, a test system consisting of the combination of a higher efficiency, lower resolution detector and a lower efficiency, higher resolution detector has been established. Source measurements have been performed to benchmark system performance and validate the accuracy of spectral simulations. Extensive simulations have been generated to establish the bounds of performance with sufficient statistical significance to determine identification performance. These results are presented in terms of detection probability and false alarm rate as a function of acquisition time for a wide variety of nuclides of interest for security applications.

Different detector combination were studied: a large slab NaI(Tl) detectors combined with HPGe detectors [2] for use in large, fixed portals and a combination of CZT detectors and CsI scintillators for use in hand-held radiometric identifiers [3].

20.2 Hybrid System for Use in Large, Fixed Portals

For the application of non-proliferation and the screening of commercial traffic, throughput must be maximized. This requires very high efficiency systems to ensure sufficient probability of detection for short acquisition times. One of the largest spectroscopic detectors is a large slab of scintillator. In this study, we consider the use of a rectangular $3'' \times 5'' \times 16''$ ($7.6 \times 12.7 \times 40.6$ cm) NaI(Tl) scintillator mounted on a photomultiplier tube and referred to here as a NaI slab. Even with good efficiency, if the system does not also have a low false alarm rate, excessive alarms require too many slow secondary measurements and searches while trying to prevent an illicit item from passing. While efficient, with more than 500 cm^2 area, the NaI slab has relatively poor resolution (7.5 % at 662 keV) and is more prone to identification failures. A large high purity germanium (HPGe) with over 50 cm^2 area and 100 % relative efficiency (compared to a $3'' \times 3''$ NaI(Tl) at 1,332 keV) is considered in this study to supplement the NaI slab and is referred to here as a HPGe100. HPGe200 indicates two such detectors. While the HPGe100 may have an order of magnitude smaller surface area and a correspondingly small solid angle, it also has about 25 times better energy resolution with a full width at half maximum (FWHM) of about 1.5 keV (0.3 %) at 662 keV. The benefit of improved resolution is not only the ability to better discern adjacent peaks, but also a reduced effective background under the much narrower peak, enhancing the signal-to-noise ratio. Figure 20.1 illustrates the detector configuration considered for this study.

The process of combining the hybrid system spectra first requires separating peak counts from scattered continuum counts. This separation is performed using standard baseline estimation. The respective continua are then separately rebinned and summed to a common energy calibration. The peak spectra are also rebinned to the common energy calibration but are further processed prior to adding to the final spectrum. The processed, hybridized peak spectra are combined with the previously

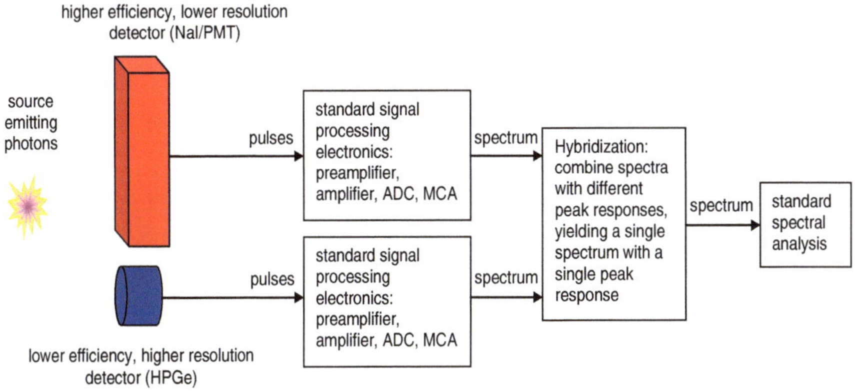

Fig. 20.1 Example hybrid detection system processing, with a large NaI(Tl) and HPGe detectors

generated continuum spectrum to produce the final output spectrum. The output spectrum has a peak shape calibration similar to the better resolution constituent detector but with peak counts from both detectors. The hybridization process only redistributes measured counts. Counts are neither created nor destroyed but are explicitly conserved. The process does not attempt to recoup scattered counts into photopeaks and so is only effective for spectroscopy detectors with significant peak responses. Since the hybrid spectrum reflects a correlation of peak responses among the constituents, all of the constituents should have similar sensitivities to be most effective across the largest dynamic range. If both detectors do not have a peak response, they will not resonate. A standard system of like detectors combines constituent spectra after matching energy calibrations. Energy calibrations might be matched by adjusting hardware settings such as amplifier gain, but variations in non-linearities and offsets limit the accuracy of this approach. More commonly, each spectrum is energy calibrated separately and is then rebinned to a common energy calibration. The rebinning process does include the introduction of some additional uncertainty. However, experience has shown that the benefits outweigh this small degradation in precision. The hybridization process also manipulates the raw spectral data, but to a greater degree effective across the largest dynamic range. If both detectors do not have a peak response, they will not resonate. A standard system of like detectors combines constituent spectra after matching energy calibrations. Energy calibrations might be matched by adjusting hardware settings such as amplifier gain, but variations in non-linearities and offsets limit the accuracy of this approach. More commonly, each spectrum is energy calibrated separately and is then rebinned to a common energy calibration. The rebinning process does include the introduction of some additional uncertainty. However, experience has shown that the benefits outweigh this small degradation in precision. The hybridization process also manipulates the raw spectral data, but to a greater degree than merely rebinning to a common calibration, incurring relatively greater additional uncertainties. One goal of this study is to demonstrate that the process of hybridization can provide a net benefit, at least for the system studied.

20.2.1 Measurements

A series of measurements were performed with a variety of sources over a range of acquisition times using the HPGe100 and NaI slab detectors. The sources included ^{241}Am (60 keV), ^{133}Ba (81, 276, 303, 356 and 384 keV), ^{137}Cs (662 keV), ^{60}Co (1,173 and 1,332 keV) and ^{152}Eu (122, 245, 344, 411, 444, 779, 867, 964, 1,086, 1,112 and 1,408 keV). These were acquired with both detectors at the same source-to-detector distances for live times of 1, 5, 30 and 60 s. Each of these live time acquisitions were repeated ten times. In addition to the individual detector spectra, associated spectra were combined to produce hybrid system spectra. Each spectrum was analyzed to obtain all of the found peak areas as well as underlying continuum backgrounds. The figure of merit for each peak was determined in accordance with

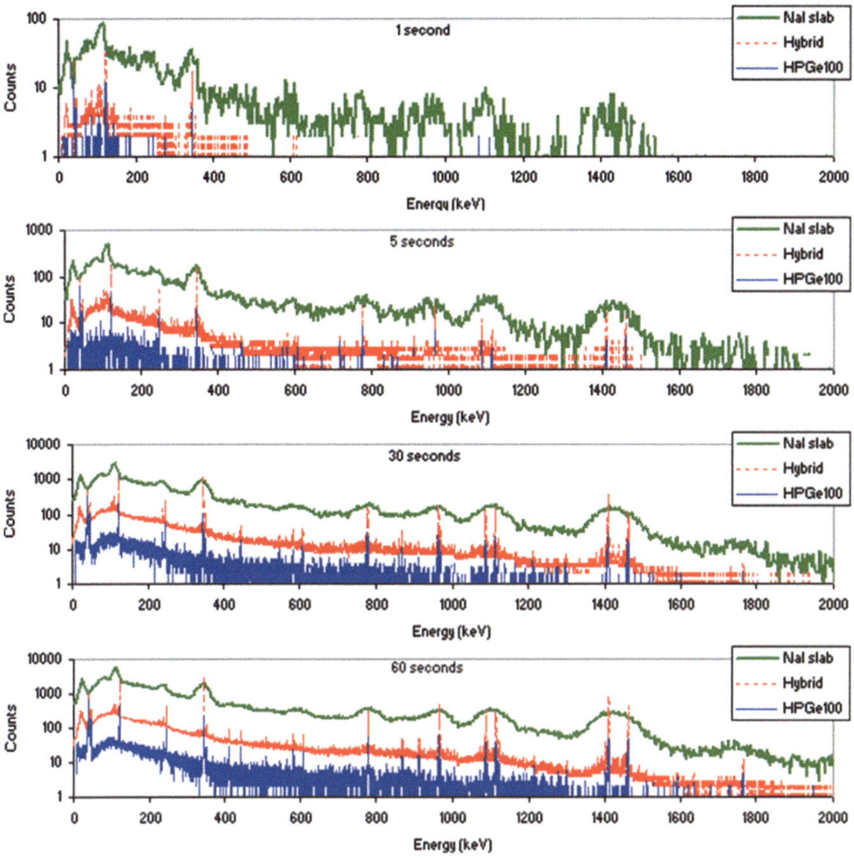

Fig. 20.2 Hybrid system measurements of ^{152}Eu for live times of 1, 5, 30, and 60 s, from *top* to *bottom*

Eq. (20.1):

$$\text{Peak Figure of Merit} = \frac{\text{Net Peak Area}}{\sqrt{\text{Peak Background}}} \tag{20.1}$$

Such figures of merit are intended for relative comparison between the significance of peaks found for the individual detectors and the hybrid combination. Figure 20.2 shows an example of the progression of ^{152}Eu spectra over the measured live times.

The measurements have also been used to benchmark simulations to provide validation of a more comprehensive assessment of a hybrid system through simulations. The simulations have been generated using response functions derived from Monte Carlo N Particle (MCNP) software [4] combined with respective background measurements.

20.2.2 Performance Evaluation

Ultimately, the performance of a security system depends on the accuracy of identification results over the full range of likely sources and acquisition times [5]. The activities are specified for point sources at a distance of 1 m. The sources include naturally occurring, medical, industrial and special nuclear materials. Each of these nuclides was modeled individually and all unique pairings, for a total of 325 simulated spectra for each detector and each live time. The acquisitions were modeled for live times of 3, 10 and 30 s. The detectors modeled were a single NaI slab, a single HPGe100 and two HPGe100 (HPGe200). Hybrid spectra were then derived by combining HPGe100 and the NaI slab (Hybrid100) and combining HPGe200 and the NaI slab (Hybrid200). The performance evaluation proceeded by analyzing each spectrum with standard nuclide identification algorithms. For the HPGe100, HPGe200, Hybrid100 and Hybrid200 spectra, the analyses used were standard Canberra Genie4 peak search, interference corrected analysis engines, with the only variation being the use of appropriate respective backgrounds for background subtraction.

The NaI slab analyses were performed using Genie's Library Correlation NID5 analysis engine, more appropriate for lower resolution detectors.

The results of all the analyses were tallied to determine the probability of detection and the false alarm rate. The probability of detection was calculated as the percentage of constituent nuclides correctly identified out of the 625 expected nuclides. The probability of detection is the complement of the false negative rate. The false alarm rate was calculated as the percentage of the 325 spectra that included any identification of nuclides that were not actual constituent nuclides, the false positive rate. The ideal response would entail identification of every constituent nuclide, 100 % probability of detection, without any extraneous identification, or 0 % false alarm rate. Depending on the specific application, one or the other of these two factors might be emphasized. The results of this performance evaluation are shown in Fig. 20.3, with the ideal location in the upper left hand corner. The plot markers for each detector configuration are the same but with increasing size to reflect increasing live times from 3 to 10 to 30 s. In general, the hybrid performance at short live times had improved detection probabilities and slightly worse false alarm rates than corresponding HPGe100 and HPGe200. At longer live times, the hybrid detection probability is about the same but the false alarm rates are somewhat better than the corresponding HPGe100 and HPGe200. One pertinent aspect of the results is the value of adding an additional HPGe detector to an existing HPGe detector compared to hybridizing with a NaI slab. At the 3 s acquisition time with the poorest statistics, both options gave similar detection probability, but the HPGe option gave a little better false alarm rate. At the 10 s acquisition time, the hybrid option provided minimal improvement while the HPGe option raised the detection probability by 15 % at a cost of about 5 % worse false alarm rate. At the 30 s acquisition time, the hybrid option yielded a minor improvement in false alarm rate but the HPGe option again raised the detection probability by 15 % at a cost of

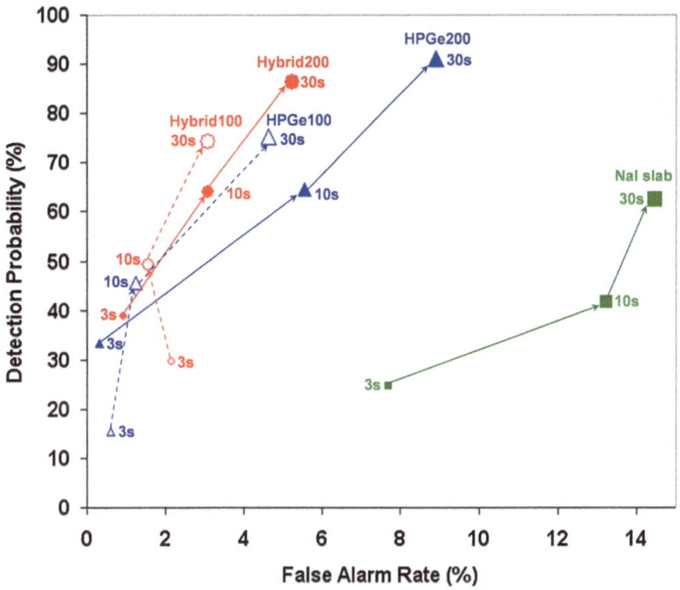

Fig. 20.3 Identification performance results for a variety of simulated detector configurations and acquisition times (3, 10 and 30 s). HPGe100 and HPGe200 are one and two 100 % relative efficiency respectively and the Hybrid100 and Hybrid200 are hybrid systems of HPGe100 and HPGe200 with a NaI slab

about 5 % worse false alarm rate. In general, the HPGe option is more beneficial, depending on the cost of extraneous false alarms for a given application. However, the addition of a NaI slab as a hybrid system proved to be a net benefit in every case. Every hybrid system performed closer to the ideal performance than the HPGe-only system, implying that contributions from the NaI slab signal more than offset accompanying noise contributions. Likewise, every NaI slab live time acquisition benefited from the addition of HPGe detector(s) spectra.

20.3 Hybrid System for Hand-Held Applications

The previous study showed the benefits of creating a hybrid system of large slab NaI detectors combined with HPGe detectors for use in large, fixed portals. Another study has evaluated the performance of hybrid combinations of CZT and scintillators for use with hand-held applications (Fig. 20.4).

Different sources (H, F, Na, Co, U, Pu ...) were simulated for live times of 3, 10, 30, 100 and 300 s. Each source energy deposition spectrum based on the Monte Carlo modeling was broadened to emulate the measured energy resolution for the respective detector type. These simulated source spectra were then combined with a live-time scaled measured background spectrum, yielding the final simulations.

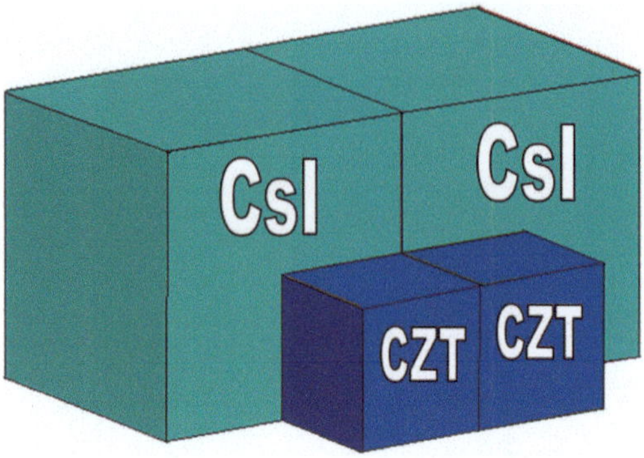

Fig. 20.4 Model of the detector configuration of the CZT/CsI hybrid system. The front face of the system is the CZT face away from the CsI

20.4 Results

The true and false identification rates were plotted for each detector configuration for each live time. The false identification rate is shown on the horizontal axis while the corresponding true identification rate is shown on the vertical axis. The ideal performance would occur in the upper left corner of the plot, with 0 % false identification rate and 100 % true identification rate. Each of the sets of points for a given detector type are connected by a line to guide the eye. The marker sizes increase with increasing live time, progressing from 3 to 10 to 30 to 100 to 300 s (Fig. 20.5).

The performance results show that the relatively high efficiency of the CsI detector enables a positive response very early, even at only 3 s. The identification performance of CsI does start to saturate at the longer counting times, with little improvement in true identification rate from 100 to 300 s. The poorer resolution of CsI results in a much higher false identification rate at every live time, with about 20–30 % worse than the CZT for most acquisition times.

The better resolution of the CZT detector does not fully compensate for the worse efficiency, without any identification made at 3 s for either the 2 or 4 cm^3 versions. The 2 cm^3 CZT also failed to make any identification at 10 s. With longer counting times, the CZT was able to approach the true identification rates of the CsI, at about 300 s, and was able to do so with much less ambiguity from false identifications.

Without some significant contributions from both detector components at 3 s, the hybrid combinations for both CZT volumes also did not make any identification at 3 s. For shorter counting times, the CsI component offers higher detection probability than the hybrid combination, although with much more uncertainty about the accuracy of the identification results. However, at almost every acquisition time the hybrid combination performed better (closer to the ideal performance in the upper left corner) than either of the separate component detectors. In general, the hybrid has higher true identification rates than the CZT component and lower false identification rates than either component.

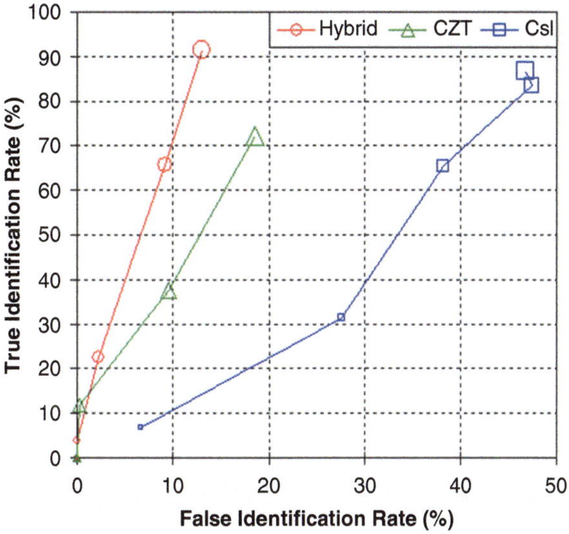

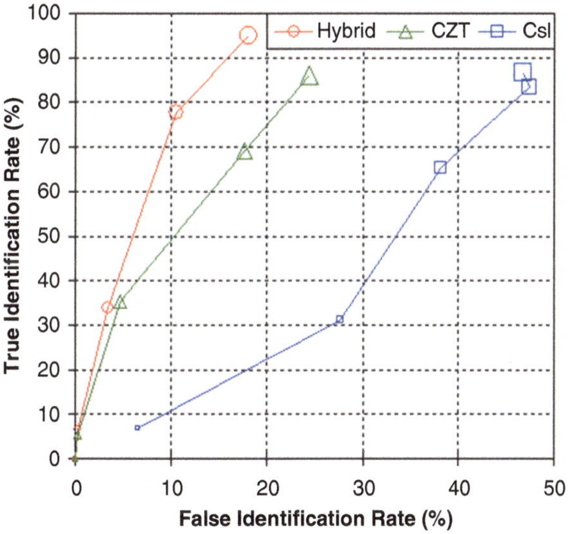

Fig. 20.5 Identification performance results for the 32.8 cm^3 CsI, 2 cm^3 CZT (*left*)/4 cm^3 CZT (*right*) and the corresponding hybrid system. Lines added to guide the eye. Marker size increases with live times (3, 10, 30, 100, 300 s)

20.5 Real-Time 3D Gamma Cartography

20.5.1 Fukushima Feed-Back and Lessons Learned

During the Fukushima accident different technologies were deployed to assess the site contamination and the dose rate level in order to define scenarios of intervention to enhance the risk management and reduce the environmental impact. Geolocalized dose rate measurements around the site, interpolation and mapping were done.

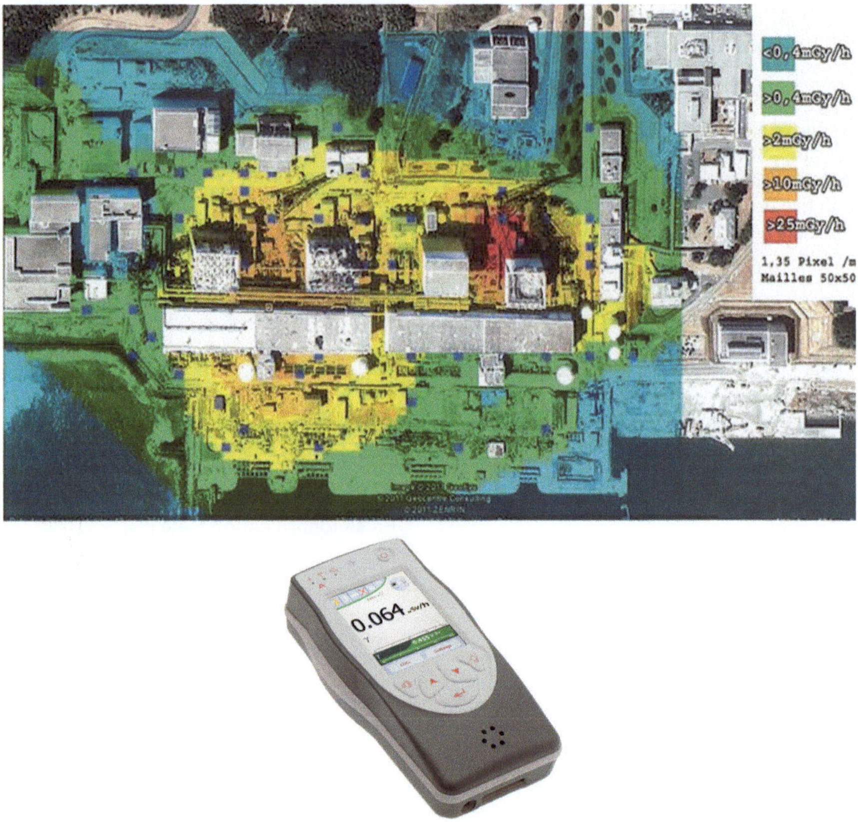

Fig. 20.6 Fukushima site dose rate mapping done with the COLIBRI platform

To execute the plan, a specific technical solution has been developed using dose mapping for preliminary assessment and post evaluation of personnel dose and to execute intervention planning as represented in Fig. 20.6. Technology used was an integrated platform developed at CANBERRA called "COLIBRI" that integrate dose rate with GPS and further data integrated with geolocalization and mapping.

From the geolocalized measurements performed with the COLIBRI rate-meter, interpolation were performed that permits first to build a map of dose rate for the entire Fukushima site and then to evaluate dose that people will encounter when they will install and operate systems. However, indoor dose rate mapping at Fukushima remains unknown for lack of GPS signal.

20.5.2 Innovation and Preliminary Results

After the Fukushima accident and due to lack of solutions, AREVA R&D teams innovated in coupling SLAM technology (used in instantaneous 3D topographic

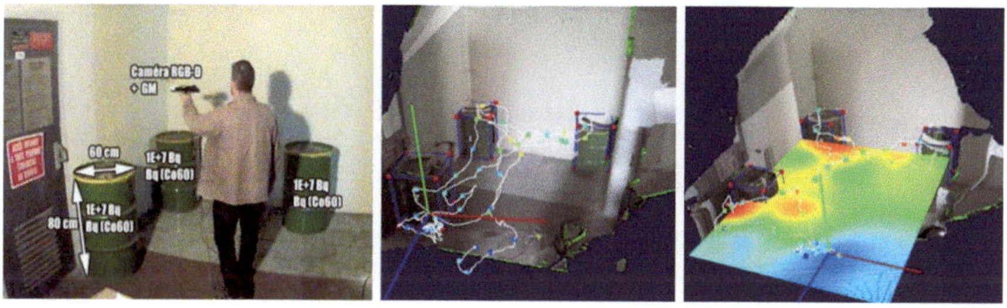

Fig. 20.7 *Left*: mapping of a cell with radioactive drums, middle: trajectory of the operator and the radiological acquisition, *right*: overlap of the 3D images and dose rate measurements

reconstruction of environment and is especially used in mining, tracking purposes, autonomous vehicles and robots positioning in volume) with dose rate and spectroscopic probes in order to develop a 3D real-time gamma cartography [6].

Today, a hand-held application is under development, the building blocks are a kinect™, a dose rate or a spectroscopic probe and a PC tablet.

Figure 20.7 shows us the different phases that are acquired simultaneously as well as the online treatment: (1) scene scanning with the kinect™ (the cameras are acquiring the geometries and the positioning in the scene), (2) radiological acquisition (dose rate or spectroscopic measurement points are acquired), (3) software modeling (in dose rate mode: an interpolation of the dose rate results on the 3D images, and in spectroscopic mode: geometrical efficiency estimation, conversion from counts/s to activity and layout of the activity on the 3D images). All these phases are performed in real time and followed by several seconds of online treatment.

This development enables us to map indoor areas, to reduce analysis time and define scenarios as quick as possible, and reduce operators' integrated dose. A drone application development is undergoing and will be dedicated for high radioactive areas and complex accessibility environment.

20.6 Conclusions

In this paper different novel nuclear measurement technologies were presented, hybrid detection was developed in order to accelerate decision in the nuclear waste domain (triage, characterization and management). Lessons learned from Fukushima helped us to innovate and develop the real time 3D gamma cartography, this breakthrough technology will be deployed soon on our sites to increase operator performances and reduce operator's dose by defining accurate intervention scenarios. Further developments were done on both technologies and they showed potential applications in security, safety and safeguards.

References

1. Russ W, Morichi M. Hybrid radiation detection system, US Patent # 7,636,638
2. Russ W, Morichi M (2009) Evaluation of a novel hybrid spectroscopic system developed at Canberra-AREVA. In: Proceedings of Global 2009, Paris, 6–11 September 2009
3. Russ W, Nakazawa D, Hau I, Morichi M (2010) CZT/NaI hybrid hand-held performance evaluation. In: Proceedings of IEEE-HST, Waltham, MA, 8–10 November 2010
4. MCNP (2005) A general Monte Carlo N-particle transport code, Version 5. Los Alamos National Laboratory report LA-UR-03-1987
5. Russ W, Kirkpatrick J, Morichi M (2006) Lanthanum bromide scintillator practical performance evaluation. In: Proceedings of INMM 47th annual meeting, Nashville
6. International Patent Application PCT/EP2014/064038 filed on 2 July 2014

Chapter 21
Cosmic Ray Generated Charged Particles for Cargo Inspection

Michael Sossong, Gary Blanpied, Sankaran Kumar, and Sean Simon

Abstract Charged particles continuously rain down on the surface of the Earth. These charged particles primarily consist of muons and electrons. Muons are subatomic particles with the same charge as the electron, but with 200 times the mass. These particles are generated from interactions of primary cosmic-rays, primarily protons, with the upper atmosphere. Decision Sciences International Corporation (DSIC) has created a tracking detector to measure the interactions of these particles with materials through which they pass: multiple Coulomb scattering and ionization energy loss and from these measurements is able to reconstruct a 3-D map of the density and atomic number of the materials in a scan volume. This map can be used to automatically detect bulk contraband (including explosives, narcotics and other materials) in the cargo as well as provide highlighting of anomalous configurations (nested or irregular volumes) for review by authorities. Fusion of the imaging with the sensitive gamma detection capability of the tracking detector enables the detection of nuclear and radiological materials even when concealed in shielding, as well as discrimination of naturally occurring radioactive materials (NORM) from nuclear and radiological threats. Times to clear most non-threat cargo range from 30 to 60 s, with suspicious scenes (heavy shielding, gamma emitting materials or materials with similar signatures to contraband materials) being held longer to confirm the presence of and identify the material. Extended scanning of suspicious scenes typically takes 2–10 min.

M. Sossong (✉) • G. Blanpied • S. Kumar • S. Simon
Decision Sciences International Corporation, Middleburg, VA, USA
e-mail: michael.sossong@decisionsciencescorp.com

S. Apikyan, D. Diamond (eds.), *Nuclear Threats and Security Challenges*,
NATO Science for Peace and Security Series B: Physics and Biophysics,
DOI 10.1007/978-94-017-9894-5_21

21.1 Introduction

The use of cosmic ray generated charged particles for the detection and identification of nuclear materials was invented at the United States Los Alamos National Laboratory (LANL).[1] DSIC licensed this technology from LANL in 2008 and commercialized it developing the Multi-Mode Passive Detection System (MMPDS). Effective inspection of the traffic at borders, airports and seaports is essential to protect the world from terrorist threats, including illegal transportation and use of nuclear materials. While the goal of the U. S. Government is to screen all traffic and cargo transported into the country, in reality only a small fraction of incoming cargo is physically inspected (5–6 % in the U.S., much lower internationally).[2] To this end, DSIC has adapted, developed and optimized for checkpoint screening, a cosmic ray particle tomography system to inspect cargo safely and efficiently. The system uses naturally occurring cosmic ray particles to obtain an image of the scanned volume, is completely passive and involves no active or additional application of ionizing or other radiation for the inspection of cargo. The design objective for MMPDS is to enable non-intrusive inspection of cargo and vehicles for undeclared goods, contraband and hidden radiological and nuclear threats using a combination of cosmic ray particle tomography and passive gamma radiation detection. Two operational systems have been built; one is located in Poway, CA, the other at the Freeport Container Port in the Bahamas. The Freeport system is fully operational and is scanning cargo containers on a regular basis. The system is currently being prepared for characterization by the Department of Homeland Security (DHS) Domestic Nuclear Detection Office (DNDO).

Decision Sciences has developed the capability to discriminate materials ranging from low density bulk contraband materials to small quantities of high-density nuclear materials. Signatures can be extracted from the 3D map providing discrimination of materials much more powerful than is available from other technologies. For example, a primary scan can clearly discriminate pallets of standard office paper from glossy paper used in magazines. This provides for the automatic detection of contraband materials from a material library as well as highlighting of anomalous configurations of materials, such as materials obscured by benign materials or expected uniform loads with irregular contents. Suspicious configurations can be reviewed by an operator to determine consistency with the cargo manifest.

[1]Borozdin et al. [1]; Morris et al. [2].

[2]Bakshi et al. [3].

21.2 Cosmic Ray Generated Charged Particles

The earth is constantly bombarded by high-energy cosmic rays that have originated from astrophysical sources. Primary cosmic rays consist mostly of hydrogen nuclei (protons) and helium nuclei (alpha particles). These primary cosmic rays interact with earth's atmosphere to produce secondary cosmic rays that are primarily pions with very short lifetimes ($\sim$30 ns). Charged pions decay into muons that have long lifetimes and are weakly interacting, therefore making it to the surface of the earth. Neutral pions decay into gamma rays that can interact with air molecules to produce electrons. While muons and electrons have the same magnitude of charge, muons are about 200 times more massive than electrons and therefore scatter less due to their higher momenta. This enables the muons to pass through most materials with very little scattering. This makes muons an effective probe for high density and high atomic number materials such as special nuclear materials (SNM) as well as dense materials typically used to shield gamma radiation. Electrons, on the other hand, are a more effective probe for differentiating lower density and lower atomic number materials due to their higher scattering and absorption in these materials.

Muons travel at relativistic speeds, are highly penetrating and can travel an average of about 20,000 m in their 60 μs mean relativistically time-dilated lifetime at their mean momentum. The muons do not arrive at the earth's surface like vertical rain, but rather with a cosine-squared angular dependence which has an average incident angle of about 37.5°. The average muon energy at sea level is 3 GeV and the flux is about 10,000/m^2/min, which increases with altitude.[3] Cosmic ray electrons (and positrons) have typical energies from 0.01 to 1 GeV and have a sea level incident flux of about 5,120/m^2/min.[4]

21.2.1 Charged Particle Signals

As the charged muons and electrons pass near a nucleus, they are subject to Coulomb scattering. The charge on the nucleus (proportional to the atomic number) and the frequency of the scattering (depends on the material density) determines the width of the distribution of scattering.

The scattering width increases with higher density and atomic number. Additionally, the charged particles lose energy to the electrons of the material through which they pass. This energy loss increases approximately with the density of the material (Fig. 21.1). By measuring the spatial distribution of the charged particle scattering and attenuation within a volume, a map of signatures related to the density and atomic number of the materials that fill the volume can be

[3] Sundaresan [4].

[4] Beringer et al. (Particle Data Group) [5].

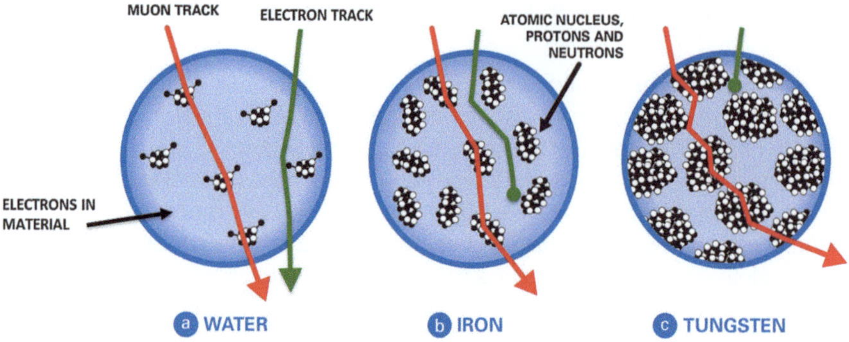

Fig. 21.1 Schematic diagram of the scattering and/or stopping of muons and electrons in different materials. Particle scattering and stopping increases as the density and the atomic number of the material increases

obtained. This tomographic map enables the detection and spatial location of threat or contraband materials by comparison with a material library. High density and high atomic number materials (including SNM) produce very strong scattering and attenuation signal. Increasing the thickness of shielding materials around SNM or other radioactive materials improves the system's ability to detect these threats using muon tomography due to the increased scattering caused by the thicker shielding. Hence, muon tomography combined with passive radiation detection is a powerful technique to detect SNM and other nuclear and radiological threats. Lower density/atomic number materials have weaker signals and more overlapping signatures, requiring more particles to provide discrimination. Because the particle flux is limited and discrimination depends on the number of particles through the material, detectable material quantities are larger for low-density materials for a fixed scan time. Interesting and threatening quantities of low-density materials are much larger, providing useful detection with scan times under two minutes. A considerable advantage of the method is that it is completely passive and relies solely on naturally-occurring incident particles rather than applied ionizing radiation.

21.3 MMPDS Hardware

DSIC has developed and manufactured several scanners of different configurations to demonstrate the capabilities of muon tomography for the detection of concealed SNM. In addition to muon tomography, the scanner uses passive gamma detection to detect concealed radioactive objects. Two large scale systems have been built, as well as two smaller systems intended for research applications. The system in Freeport (referenced earlier), is capable of scanning full-size trucks and shipping containers (40 ft and larger) for concealed SNM materials. The second system,

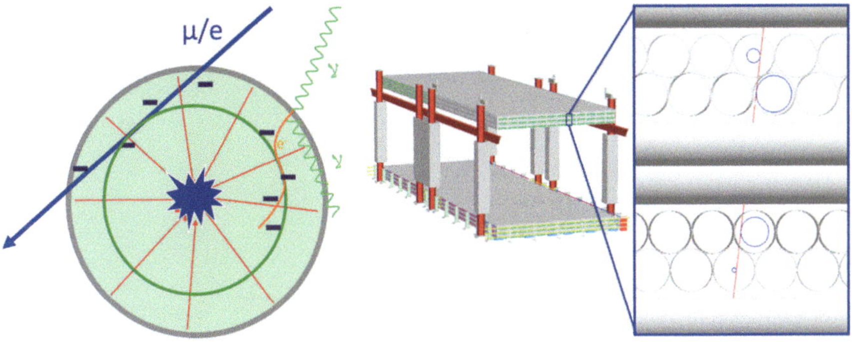

Fig. 21.2 Sealed drift tubes function by detecting ionization caused by the passage of charged particles through the gas volume. Using timing information of the arrival time of the ionized electrons, a radius can be calculated. This provides a sub-millimeter position resolution measurement of the path of the charged particle. The 3-D trajectory of the particle is reconstructed from measurements in several layers of tubes

based at the DSIC facility in Poway, CA, is used for testing and optimization of the detector. The two research systems were delivered to commercial customers.

The sealed drift tube is used as the primary sensor element in the scanner (Fig. 21.2).[5] A drift tube is an ionization detector that produces electrical signals in response to ionizing particles or radiation that pass into or through its volume. Sealed drift tubes are very robust, durable sensors requiring no maintenance over many years. Each tube can provide high-precision position measurement of a traversing charged particle over a large area while requiring only one signal processing channel. It is a sealed, gas-filled cylinder that has conducting walls (cathode) and a fine wire element strung longitudinally down the tube (anode). A high voltage is applied between the anode and the cathode. The gas in the drift tube mixture is ionized by the passage of muons or electrons which results in a number of electron-ion pairs (on the order of 25 pairs per 1 cm path of a muon). The drift-tubes comprising the scanner are also capable of detecting gamma rays. The incidence of gamma rays on the aluminum walls causes Compton electrons to be emitted from the conductive wall of the tube, causing ionization of the drift tube gas.

The drift tube electrical signals are detected, amplified and identified by first stage electronics. The temporal information of the electrical signal can be used to determine the closest approach radius between the charged particle path and the wire. Custom electronics, installed at one end of the tubes, acquire, time stamp and filter tube signals. Data is delivered to an analysis cluster running software to identify muon and gamma events, calculate tracks and perform tomographic reconstruction that produces a 3-D map of the materials of the volume under

[5]For a review of ionization chamber detectors, please refer to Chapter 5, Knoll [6].

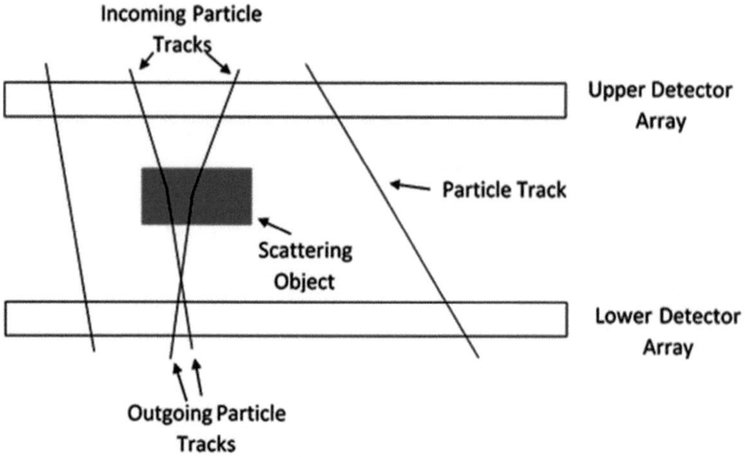

Fig. 21.3 Schematic diagram showing muons traversing the detector array and scattering in an object in the scan volume between the detector arrays. The scattering angle is greatly exaggerated in this figure; the actual angle is a few milliradians

inspection. Count rates on each tube are individually analyzed to produce a map of the position and distribution of radioactive sources within the volume.

The scanner is constructed using large arrays of sealed drift tubes, arranged in orthogonal layers, to enable 3-D tracking of muons and electrons. These arrays are arranged above and below the volume of interest and used to track each charged particle as it enters and leaves the detector (Fig. 21.3). Changes in the trajectory of the particle are analyzed to produce a 3-D representation of the materials in the volume. If the particle is attenuated in the volume, the absence of an outgoing trajectory is recognized and analyzed for the attenuation image.

21.4 Imaging and Detection Software and Algorithms

In the DSIC scanner, cosmic ray muons and electrons are tracked into and out of a volume of interest. Their collective scattering and attenuation information is then used to reconstruct the materials through which the muon and electrons have passed. Each scattered particle provides a measurement of the scattering angle and an approximate location of the scattering. These data are used to reconstruct a 3-D map of signatures related to the proton and electron densities of the interrogated materials. After tomographic reconstruction from particle scattering and momentum data, threat objects may be identified and distinguished from benign cargo on the basis of their size, shape, atomic number and density. For SNM and other high-Z, high density nuclear threats, muons are the primary probe, whereas for lower density contraband, including explosives and narcotics, a combination of muon and electron dynamics is most effective in detection and discrimination of the materials.

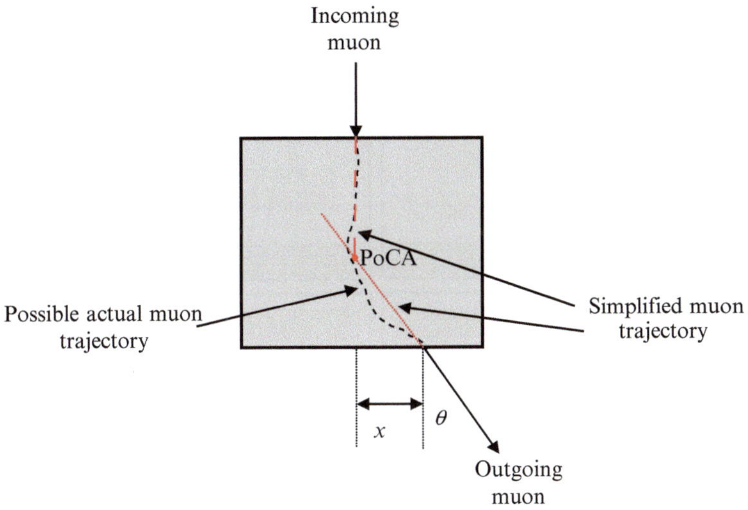

Fig. 21.4 Schematic diagram of the scattering of a cosmic ray particle in an object. The actual particle trajectory inside the object is not known, but the incoming and outgoing trajectory of the particle can be measured. The scattering angle is the angle θ between the incoming and outgoing particle trajectories. In a simplified model, the scattering location is considered to be in the region where the extrapolated incoming and outgoing trajectories are closest to each other (they will not necessarily intersect as shown in the figure). This location of closest approach is called the 'point of closest approach' or 'PoCA'

21.4.1 Charged Particle Imaging

The image reconstruction starts with determining the incoming and outgoing tracks of the muons and electrons as they pass through the top tracker, the scanned volume and the bottom tracker. The tracks are determined using the locations on the drift tubes at which the muons and electrons were incident. As stated above, this positional information can be derived from the temporal information of the electrical signal in the output of the drift tube (Figs. 21.4 and 21.5).

21.4.2 Multiple Coulomb Scattering Imaging

The scan volume between the detector arrays is divided into voxels.[6] Once the tracks are determined, the scattering angle θ and the scattering location and its corresponding voxel are estimated. This estimation varies based on the reconstruction algorithm used. A distribution is accumulated for each voxel. As more muons

[6]For a description of a reconstruction method, please refer to Schultz [7].

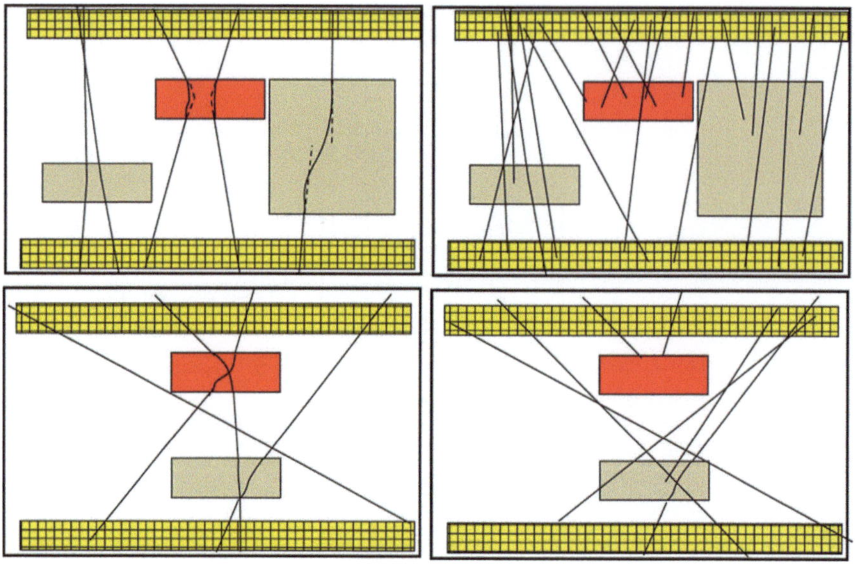

Fig. 21.5 Schematic diagram of scattering and stopping of tracks passing through objects in the scan volume between the upper and lower detector arrays (*colored yellow*). The scattering angle reflects the integrated proton density through which the particle passes. The scattering point provides information as to the vertical location of the scattering source. The distance of closest approach between the reconstructed trajectories provides information related to the physical thickness of the material traversed. Attenuated particles provide information related to the integrated stopping power of the material through which they pass. For both techniques, multiple angle exploration of the volume by charged particles helps to resolve remaining ambiguity as to the position of materials in the volume

and electrons enter and scatter in the scan volume, the voxel scattering strength distribution is updated. The scattering density map is calculated at the end of a scan period using a statistic characterizing the mean square scattering of the distribution of scattering per unit depth within each voxel. Iterations of this process produce better estimates of the portion of scattering caused by each voxel and therefore higher fidelity estimation of the scattering density along the path of the charged particle.

21.4.3 Charged Particle Attenuation Imaging

In addition to the scattering information, the attenuation of the muons and electrons is used to help reconstruct the scan volume. The attenuation of the particles inside the scan volume results in a particle track being detected at the upper detector array without a corresponding particle track being detected at the lower detector array. The momentum loss density map is calculated at the end of a scan period using a statistic characterizing the mean momentum loss per unit depth within each voxel. Iterations of this process produce better estimates of the portion of energy loss caused by each traversed voxel, therefore producing higher fidelity estimation of the momentum loss along the path of the charged particle.

21.4.4 Gamma Detection and Localization

Smuggling of nuclear material is very likely to involve an attempt at hiding the gamma emissions of the nuclear material using a high density shielding material. Shielded packages containing nuclear and radioactive materials can be imaged from their scattering and absorption characteristics of charged particles. If the material is transported in an unshielded configuration, gamma emissions allow the detection and location of nuclear or radiological materials. As stated in Sect. 21.2, the drift tubes detect gamma radiation. While individual drift tube sensitivities to incident gamma radiation are low, the large field of regard for the assembled arrays results in a highly sensitive passive gamma detector. The arrays are arranged layers of drift tubes, providing greater than 30 % efficiency for detection of a gamma ray incident on the detector array. In order to accurately determine the presence of a gamma source in the scan volume, the background level of gamma radiation needs to be measured and accounted for. Since this background level can change during the day, the background needs to be monitored periodically. This is accommodated in the MMPDS through an automated calibration process. Additional corrections can be made for attenuation of background gamma rates in the object under inspection. The presence of multiple sensors at known locations enables generation of a spatial map of the radiation intensity which enables the source to be spatially localized and facilitates differentiation of NORM from threats (which tend to be point sources).

21.4.5 Nuclear and Radiological Material Detection

The DSIC MMPDS currently uses a fixed portal configuration that integrates cosmic ray (electron and muon) scanning with passive gamma detection. These stationary scanners employ two detector arrays (one array above and another below the scanned volume) to track incoming/outgoing muons and electrons and sense gamma emissions. The conveyance is moved into the scan volume and scanning commences once it is stationary. Charged particle imaging and gamma detection are performed concurrently during data acquisition. At periodic exposure times, the gamma emission signal and material map are evaluated for the presence of threat materials. When no significant gamma emissions are detected, the material map is searched for thicknesses of shielding that could be blocking these gamma emissions from a specified minimum SNM mass or radiological material strength. If no such shielding is found, the conveyance is cleared, with a target scan time of

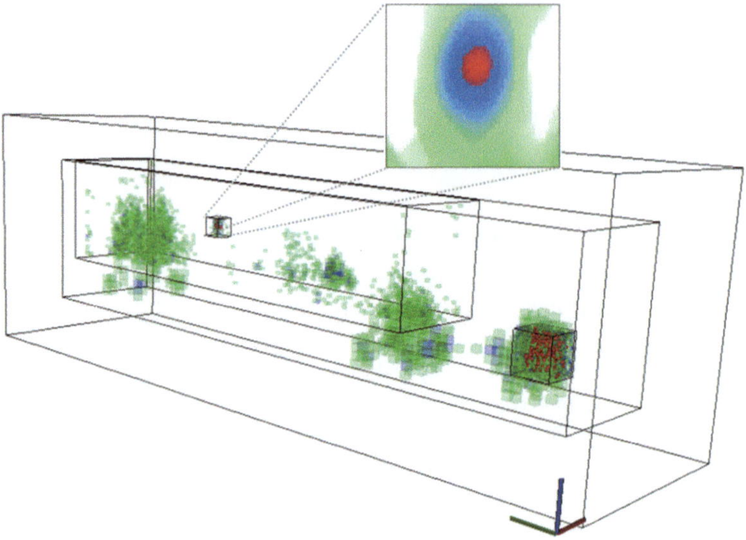

Fig. 21.6 Reconstructed map of the scan volume showing high scattering regions in *red*. A threat has been detected and located and the corresponding region has been magnified and shown

less than 60 s. In the case where gamma emissions are measured, the material map is searched for the specified minimum SNM mass. The gamma location information can be used to guide this search. If no such mass is found and the gamma levels are determined to be below an established threat level, the conveyance is cleared. In the case where sufficient shielding is identified in the material map, extended scanning is employed to confirm or refute the presence of SNM within the identified shielding region (Fig. 21.6). If no such mass is found within the shielding, the conveyance can be cleared. If the presence of high-Z SNM is confirmed, the conveyance is referred to the appropriate response authorities. MMPDS performs this analysis automatically for primary scanning, providing the operator with either a go (clear) or no-go (suspicious) indication.

21.4.6 Detection of Contraband

Decision Sciences has developed the capability to automatically detect bulk contraband such as drugs and explosives as well as highlight suspicious configurations of materials in a cargo container. Several materials, including both threats and confounding materials (materials with signatures similar to threat materials) have been scanned. From this data, a preliminary library has been populated with both threat and benign materials (Fig. 21.7). Independent scans of these materials have

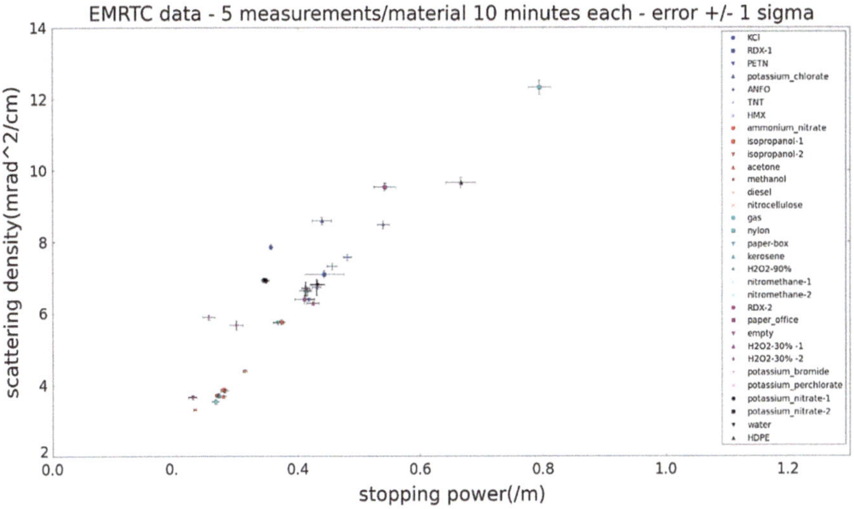

Fig. 21.7 Material discrimination based on measured cosmic ray parameters of scattering density and stopping power based on a 10 min scan of a number of materials. The one standard deviation error bars are also shown. The parameters are measured from the image of the volume and can be used to differentiate and identify materials. The plot shows a number of lower density materials. Metals and other higher density materials can also be differentiated with their parameters having a significantly larger value that the lower density materials that are shown on the plot

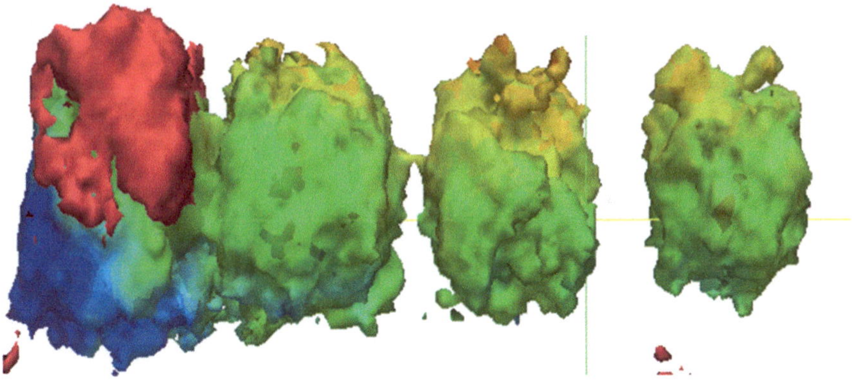

Fig. 21.8 Cosmic ray scan image (3 min exposure) of four pallets of office paper with glossy paper replacing the top layer on the left-most pallet. Paper coloring is based on the uncorrected scattering signal. The volume segmented and identified as glossy magazines is false colored red according to its signature match

demonstrated the capability to automatically detect and identify the materials of interest (Fig. 21.8). Current development is focused on expanding the library of materials and enhancing performance for complex clutter situations.

21.5 Testing and Evaluation

In August of 2011, Decision Sciences contracted the operator of the Nevada Nuclear Security Site (NNSS), National Security Technologies, LLC (NSTec), to perform tests of a laboratory prototype MMPDS for SNM detection. During 10 days of testing, the system demonstrated zero downtime. More than 1,800 scans were performed using three different SNM masses, ten lead and steel shielding configurations and a collection of cargo and clutter ranging from standard cargo to very heavy clutter and NORM. System performance was characterized. A great deal was learned about the system's operations and capability to detect nuclear materials. Decision Sciences continues to development new features and functions as well as enhance the performance of the system.

Decision Sciences is currently under contract with the U.S. Department of Homeland Security (DHS) Domestic Nuclear Detection Office (DNDO) and is preparing for system characterization by a government test team.

Decision Sciences has successfully completed a contract with the Department of Defense (DOD) Terrorism Support Working Group (TSWG) Combating Terrorism Technical Support Office (CTTSO) to evaluate the capability (in simulation) to identify explosive materials. Additional development testing has been performed internally and independent demonstration and testing is being planned for 2015.

21.6 Commercialization and Deployment

21.6.1 MMPDS Operational at Freeport Container Port, Bahamas

The first article laboratory prototype tractor-trailer scanner was completed in Poway, California in 2011. Construction began in February, 2012, of the tractor-trailer scanner at an operational port – Freeport Container Port located in the Bahamas. The first system demonstration occurred in August, 2012. The system is installed at the entry/exit gate of the container port and scans all inbound and outbound traffic (Fig. 21.9). Since commissioning the detector, Decision Sciences has acquired data through scans of almost 2,000 containers. Valuable commerce data is being cataloged.

21.6.2 Port Scanning Operational Flexibility

The tractor-trailer MMPDS installed at the Freeport Container Port is a convenient design for scanning containers loaded on tractor-trailers. It can be installed at existing interchange islands and scanning is performed concurrent with other activities.

Fig. 21.9 Pictures of the Freeport installation of the MMPDS. The scanner has been in operation since August 2012. Valuable commerce data is being collected and analyzed

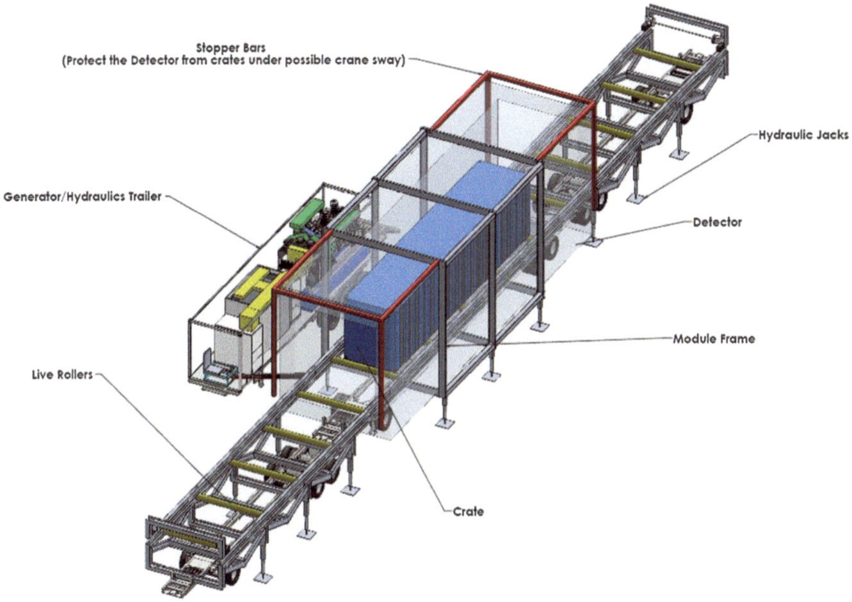

Fig. 21.10 Container handling configuration design drawing. A configuration such as this would better accommodate on-pier scanning at a port for transshipped containers. A smaller version of this configuration could accommodate the scanning of air cargo containers in line with aircraft loading

Other configurations may be more appropriate for performing on-pier scanning at a port for transshipped containers. Decision Sciences is currently designing a relocatable MMPDS that handles and scans the containers delivered and retrieved by cranes or straddle carriers (Fig. 21.10). Such a system can be scaled according to throughput and port operational concept to provide cost effective architecture for both high and low volume scanning. The container handling configuration of the MMPDS can also be scaled to accommodate air cargo scanning or pallet scanning in-line with aircraft loading or at a warehouse or storage location.

21.6.3 Other Nuclear Security and Safety Applications for Cosmic Ray Scanning

Key properties of cosmic ray charged particle scanning including the high signal from nuclear materials relative to common materials, the penetration of very heavy clutter and shielding and absence of added radiation, make it a viable candidate for many scanning and monitoring operations in the nuclear industry. For example, muon scanning can be used to scan or monitor fuel rods within storage vessels. Muons can also be used to scan nuclear reactor cores to confirm the presence of the fuel as well as determine its configuration in the case of a meltdown or other incident.[7] Decision Sciences signed a contract with Toshiba Corporation on August 8, 2014 to implement a scanning system to aid in the restoration of the Fukushima Daiichi Reactor complex. DSIC will design, manufacture and deliver a detector and tube arrays that fit into the Daiichi power plant building. The detector will be part of Toshiba's overall effort at Fukushima to determine the location and condition of the nuclear fuel inside the reactor buildings. The information provided by the detectors will assist Toshiba in developing a safe and effective remediation plan. This application assists the restoration project in two ways – detects and identifies the nuclear material and supports and secures a safe working environment for personnel. Tokyo Electric Power Company (TEPCO) estimates that Muon Tomography (MT) can shorten Fukushima restoration efforts by 10 years.

21.7 Summary

Cosmic ray scanning is the only passive non-intrusive imaging solution. Decision Sciences has commercialized cosmic ray scanning for detection of bulk contraband and nuclear and radiological materials in maritime cargo containers and occupied vehicles. A prototype has been installed and is operational at the Freeport Container Port in Freeport, Bahamas. The technology may have many other applications for enhancing nuclear security. The reader is encouraged to contact the authors to discuss potential applications.

References

1. Borozdin KN et al (2003) Surveillance: radiographic imaging with cosmic-ray muons, Nature 422:277. http://dx.doi.org/10.1038/422277a
2. Morris CL et al (2008) Tomographic imaging with cosmic ray muons. Sci Glob Secur 16(1–2):37–53

[7] John Perry et al. [8].

3. Bakshi, Flynn SE, Gans N (2011) Estimating the operational impact of container inspections at international ports. Manag Sci 57(1):1–20
4. Sundaresan MK (2001) Handbook of particle physics. CRC Press, Boca Raton
5. Beringer J, Particle Data Group et al (2012) Review of particle physics. Phys Rev D86:010001
6. Knoll GF (1979) Radiation detection and measurement. Wiley, New York
7. Schultz LJ et al (2007) Statistical reconstruction for cosmic ray muon tomography. IEEE Trans Image Process 16(8):1985–1993
8. Perry J, Azzouz M, Bacon J, Borozdin K, Chen E et al (2013) Imaging a nuclear reactor using cosmic ray muons. J Appl Phys 113:184909. doi:10.1063/1.4804660

Part IV
Nuclear and Radiological Preparedness

Chapter 22
Lawrence Livermore National Laboratory's Contribution to U.S. Preparedness for Nuclear and Radiological Threats

Bruce E. Warner and Karen Rath

Abstract This talk highlights a few of Lawrence Livermore National Laboratory's many contributions to U.S. nuclear and radiological threat preparedness. The Laboratory is part of a larger partnership between the DOE/NNSA, the national laboratories, and other agencies of the U.S. government to support emergency response and consequence management for nuclear and radiological incidents. Lawrence Livermore has many decades of experience in understanding the threat posed by the full range of weapons of mass destruction worldwide, including chemical, biological, nuclear, and radiological. At the other end of the preparedness spectrum, the Laboratory has a long-standing role in support of U.S. nuclear and radiological incident response including atmospheric transport and dispersion modeling and nuclear forensics assessments and laboratory analyses of special nuclear and other radiological materials. We staff incident response teams, support planning and preparedness, and provide first responder training and proficiency assessments.

The combination of being threat informed and operationally involved provides valuable insight into the research and development needs of the U.S. for preparedness. The Laboratory is accomplished in the development of radiation detection equipment and systems. Recent advances include the development of novel radiation detection materials, such as a plastic scintillator that can differentiate neutrons and gamma rays that could be used in portal monitors. Lawrence Livermore also supports U.S. nonproliferation and treaty verification goals with technical and operation support for the International Atomic Energy Agency (IAEA) and the Comprehensive Nuclear-Test-Ban Treaty Organization.

B.E. Warner (✉) • K. Rath
Lawrence Livermore National Laboratory, Livermore, CA, USA
e-mail: warner2@llnl.gov

S. Apikyan, D. Diamond (eds.), *Nuclear Threats and Security Challenges*, NATO Science for Peace and Security Series B: Physics and Biophysics,
DOI 10.1007/978-94-017-9894-5_22

22.1 Nuclear Security Is at the Laboratory's Core

Lawrence Livermore applies the science, technology, and engineering expertise developed to ensure the safety, reliability, and security of the U.S. nuclear stockpile to make the nation safer. We work to develop solutions that can be used protect against a broad range of threats and address the global security issues of nuclear nonproliferation, preparedness, prevention, protections, and response and recovery.

Nuclear and radiological threats have grown over time, starting with the Cold War, where large nation states had thousands of military warheads. The proliferation threat continued to grow as additional nation states developed nuclear weapon arsenals. These threats still exist today, but now there is the added threat of nonstate actors who are interested in mass disruption by any means possible including nuclear and radiological (i.e., a dirty bomb).

The Laboratory provides the intelligence community and U.S. Armed Forces with scientific and engineering expertise in fields relevant to the development, design, engineering, manufacture, and delivery of chemical, biological, radiological and nuclear weapons. We provide information about critical processing steps or production facilities that could be used for the development and manufacturing of weapons of mass destruction.

We support the U.S. government with contributions that address the full cycle of threat device development, transport, and use. We are concerned with planning, acquisition of material, design, and manufacturing. The Laboratory also develops technology to support the multiple government agencies that would be required to respond to the crisis if a weapon were used and to restore critical U.S. facilities as soon as possible.

In this paper we discuss Laboratory contributions to radiation detector materials, the National Atmospheric Release Advisory Center (NARAC) and the Forensic Science Center.

22.2 Radiation Detectors

Detectors are a critical component of preparedness for nuclear and radiological threats, and the Laboratory continues to drive development of the science and engineering to detect and identify illicit sources of plutonium and uranium. Both plutonium and highly enriched uranium are typically identified using a combination of devices, working together, to detect their invisible gamma and neutron radiation emissions. Since 9/11, there has been a renewed drive to search for more definitive radiation detection and identification technologies.

Lawrence Livermore has developed sophisticated tools to detect and interdict nuclear materials and continues to improve detector performance, reduce costs, reduce the need for scarce materials, and resolve operational constraints. This is especially important for field use, where radiation detectors must be small, robust, operate at ambient temperature, and provide high-detection efficiency.

22.3 High-Purity Germanium Detectors

Detectors made of high-purity germanium, a semiconductor, have long offered the best energy resolution, allowing precise identification of the gamma rays emitted by plutonium and uranium. Lawrence Livermore has continued to improve these detectors by combining methods so that they can both identify and locate sources. The Laboratory's GeMini was originally developed for NASA's Voyager space mission and was further developed to become an R&D 100 Award-winning portable detection device.

GeMini features a detector made from an ultrapure germanium crystal. Many other substances can be used to detect gamma rays, but germanium offers the best resolution. However, germanium achieves its spectacular resolution only when cooled to cryogenic temperatures of about 100 K (−173 °C) or less. GeMini's design incorporates an innovative ultraminiature cooling system, rugged construction, low power consumption, automated operation, and small size. In fact, the instrument is small enough to fit in the palm of one's hand.

GeMini's outstanding energy resolution is particularly important in national security applications, when it is critical to differentiate between legitimate and illicit sources of gamma rays. The portability of GeMini makes it ideally suited for first responders either in search operations or in the case of a natural disaster where a concern of radioactive contamination exists. A version of GeMini is being built for the international safeguards community to use in field inspections of nuclear processing facilities.

22.4 New Sensing Materials

An alternative method for detecting gamma rays is scintillation, in which radiation interacts with a material to produce a brief but measurable flash of light. Although high-purity germanium, a semiconductor, has been the standard for years, it is expensive and hard to scale for large applications such as border crossing detection. These limitations drove the Department of Homeland Security's request to develop more effective materials to detect gamma rays. In response, Livermore essentially went back to first principles to develop new materials for cheaper, faster, and more accurate sensors.

Our new scintillating materials include single-crystal, plastic, and ceramic materials with improved radioisotope identifiers. These materials offer high-energy resolution performance, can be much less costly, and may be more easily fieldable. Furthermore, they are easy to manufacture and solve key technical problems.

22.5 Single-Crystal Materials

One of Lawrence Livermore's new R&D100 Award-winning radiation detection materials is made from strontium iodide doped with europium, $SrI_2(Eu)$. Radiation interacts with this scintillating material, producing a measurable flash of light when exposed to plutonium or highly enriched uranium. The precision of the scintillator material's response, or energy resolution, defines the material's ability to distinguish between gamma rays that have similar energies. $SrI_2(Eu)$ is an easily grown crystal, has excellent energy resolution, produces more photons than other comparable materials, and is not radioactive. Small crystals have demonstrated 2.5-percent energy resolution (Fig. 22.1).

Fig. 22.1 (**a**) A crystal of strontium iodide doped with europium for gamma-ray detection is an R&D 100 Award winner. (**b**) The crystal is encapsulated when used in a radiation detection device and (**c**) glows *blue* when exposed to ultraviolet light

The $SrI_2(Eu)$ scintillator can potentially serve a wide range of applications that use gamma-ray spectroscopy to identify radioisotopes and can be easily incorporated into the handheld radiation detectors being produced by many companies.

22.6 Plastics

Although highly sensitive and reliable detection systems are available today, existing systems consist of certain materials that are difficult to deploy in large volumes. What's needed is a new material that is easily fabricated and deployed. An effective detector in portal-inspection applications must also differentiate potentially dangerous radioactive sources from nonthreatening sources, such as cosmic rays, fertilizers, ceramics, decorative uranium glassware, welding rods, medical radioisotopes administered to cancer patients, and even bananas.

Lawrence Livermore collaborated with Eljen Technology in Texas to develop an enhanced plastic scintillator material that is a solid solution to efficient neutron and gamma-ray differentiation. The material consists of a very high concentration of scintillating dye suspended in a polyvinyltoluene polymer matrix. Eljen Technology, Livermore's industrial partner, commercialized the new formulation and has a product on the market.

Plastic scintillators pose none of the hazards or difficulties associated with organic crystals or liquid scintillator options. Given the low cost of plastic scintillator material, the material could be economically fabricated onto far larger surface areas than is possible with other neutron-detector materials, improving the nation's ability to identify radiation from illicit sources at transportation portals.

22.7 Using Software to Enhance Detection

Laboratory-developed software applies advanced algorithms and machine-learning methods to extract more information from available signatures at border-inspection points. By applying these methods, we are able to reduce false alarm rates significantly and improve sensitivity to signatures we are concerned about.

22.8 National Atmospheric Release Advisory Center

Another key capability in place at Lawrence Livermore to support crisis response is the National Atmospheric Release Advisory Center (NARAC), which provides tools and services to predict and analyze the spread of hazardous materials in the atmosphere. Established in 1979 after the Three Mile Island nuclear power plant accident, NARAC has continued to respond to emergencies involving nuclear, radiological, chemical, biological, and hazardous natural emissions into the atmosphere. The center is available 24 h a day, 7 days a week to conduct emergency planning, real-time assessment, emergency response, and detailed studies of incidents.

NARAC is continuously ready to respond to an emergency: every day the center receives more than a million meteorological real-time observations and weather forecast data feeds from the National Weather Service's National Centers for Environmental Prediction, the U.S. Navy and Air Force, and many other sources around the world. This detailed meteorological information is combined with comprehensive databases of maps, terrain, land use, population, material properties, and release mechanisms. The result is that NARAC's operations staff can use computer models to generate predictions that focus on almost any area in the world at an urban, regional, or global scale.

During an event, NARAC staff members work closely with field monitoring teams, emergency responders, operations centers, national laboratory experts, and federal, state, and local government agencies engaged in the response. Timely and accurate NARAC plume predictions aid emergency preparedness response efforts to protect the public and the environment.

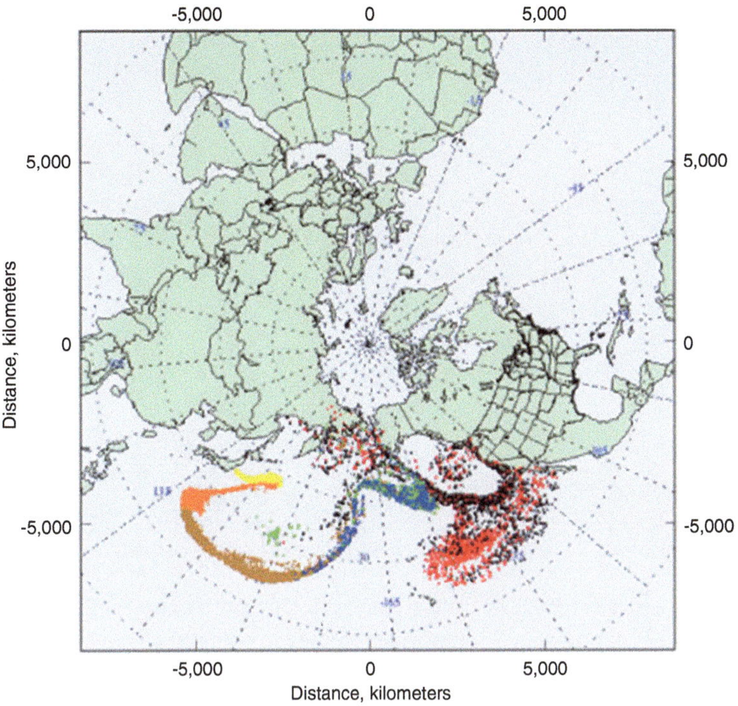

After the Fukushima Daiichi nuclear power plant accident in Japan, NARAC provided regular forecasts to support mission planning, estimates of possible dose in Japan, and predictions of possible arrival times and dose in U.S. territories. NARAC simulations of total external dose rate show combined effects of airborne and ground contamination despite the rapidly changing meteorological conditions that presented a significant modeling challenge. NARAC generated hypothetical scenarios, reconstructed events, and refined estimates of plumes based on atmospheric dispersion modeling.

22.9 Forensics: Capabilities for the Full Range of Threats

Livermore scientists play a variety of key roles in fostering international cooperation to help track and control nuclear materials. Our nuclear forensics investigators search for evidence that will help them determine the history and source of nuclear or radioactive materials found outside regulatory controls. Lawrence Livermore has honed its forensics abilities on real-world samples and has analyzed every interdicted sample since 1994.

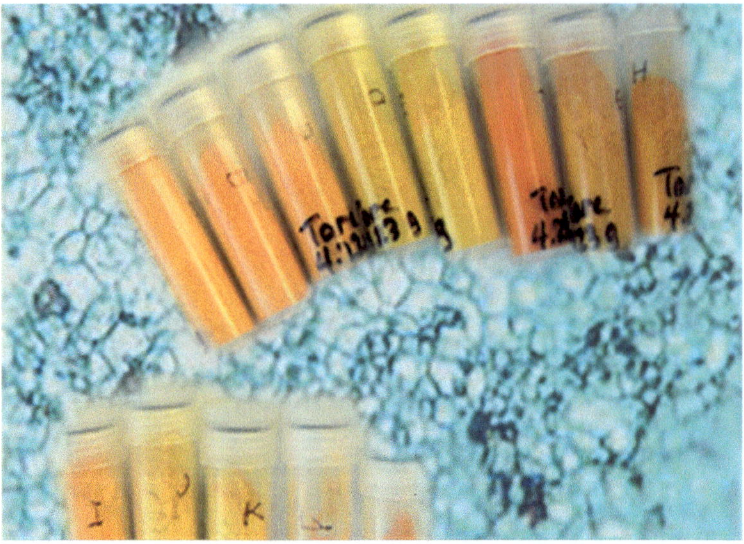

A major security concern of the international community is the activity of groups, whether nation-states or terrorists, seeking to obtain such materials for illicit purposes. To address this threat, several national and international endeavors seek to control the spread of nuclear materials as well as the technology and expertise associated with their production and use.

Lawrence Livermore is an active participant in these efforts, bringing its many decades of expertise in nuclear weapons design and performance to the newer discipline of nuclear forensics. In nuclear forensics, investigators act as "sleuths," analyzing nuclear or radioactive materials for clues to a material's source to identify where legitimate, legal control was likely lost.

One tool is the Livermore-developed Uranium Sourcing Database and the innovative database query system DAVE (Discriminant Analysis and Verification Engine), which has contributed significantly to the Laboratory's efforts to foster international cooperation in nuclear forensics. The Uranium Sourcing Database now contains approximately 190,000 data points, representing more than 6,300 samples from 133 distinct sources and 31 different countries.

The combination of the Uranium Sourcing Database and DAVE allows Livermore scientists and others to determine the origin of unknown materials sometimes seized on the other side of the globe by police or border security. As part of DOE's international outreach efforts in nuclear forensics, the Laboratory shares the database structure, lessons learned, and the analysis algorithm used in DAVE. Armed with this information, individual countries can build country-specific databases and query systems, creating one of the key capabilities of their own nuclear forensics programs. The goal of the U.S. is to encourage "capacity building," that is, improving the ability of other countries to conduct nuclear forensics and, at a minimum, be able to detect materials that are outside regulatory controls.

Building international communities is key to creating a nuclear-secure future. These international issues require global engagement to address the critical twenty-first-century problem of illicit tracking of nuclear materials. Lawrence Livermore's nuclear forensic experts share their knowledge through multilateral and bilateral efforts, workshops, database technology, and direct scientist-to-scientist interactions.

22.10 Tools to Help the U.S. Government Address Nuclear Threats

Livermore Laboratory draws on expertise developed for the stockpile stewardship mission and applies it to address dangers from nuclear proliferation and terrorism. We provide science and engineering breakthroughs in threat-informed big-data analytic tools, detection of nuclear and radiological materials, real-time crisis response to nuclear incidents and accidents, and forensics to track and control nuclear materials. These breakthroughs help protect the U.S. from radiological threats and help ensure that the nation is ready in the event of a nuclear incident.

For more information, see *Science and Technology Review* articles, "A Solid Solution for Neutron and Gamma-Ray Differentiation" (https://str.llnl.gov/OctNov12/zaitseva.html), "A Scintillating Radiation Detection Material" (https://str.llnl.gov/OctNov10/cherepy.html), "Livermore Responds to Crisis in Post-Earthquake Japan" (https://str.llnl.gov/JanFeb12/sugiyama.html), "Promoting International Security through Nuclear Forensics" (https://str.llnl.gov/October-2014/ramon), and Identifying the Source of Stolen Nuclear Materials" (https://str.llnl.gov/str/JanFeb07/Smith.html).

Chapter 23
The Emerging Threat Environment and the Impact on Nuclear Preparedness

Vayl Oxford

Abstract The emerging threat environment is dominated by several critical factors that will have significant consequences on the global security environment along with commensurate impacts on nuclear preparedness. The combination of dissolution of nation states into culturally and ethnically aligned entities, the reduction of military presence in the Middle East resulting from the withdrawal of troops from Iraq and Afghanistan, the likelihood of Iran becoming a nuclear weapon state and the recent Russian aggression in Crimea and Ukraine give rise to a geopolitical and associated threat environment that is unprecedented in recent history. This paper explores these factors and identifies some the potential security impacts as well as the likely consequences or challenges to nuclear security and preparedness. Lastly, proposed next steps for the U.S. and NATO/Europe are provided as a means to initiate a dialogue to protect against and mitigate the effects generated by these global factors.

23.1 Introduction

Preparing for a low probability, high consequence event is a complex proposition especially for events involving nuclear weapons. Yet, Americans and people around the world impacted by the attacks of September 11, 2001 and those charged with responsibility to react to the post-9/11 threat environment will recall the sense of urgency and unity of effort that went into preventing future attacks. Today, preparing is even more complicated due to the fading memory of the trauma and vulnerability felt on 9/11 and is being replaced by a sense of complacency that is not only influenced by the time since the 9/11 attacks but also by the lack of any subsequent attacks. Moreover, a growing number of the world's population was either very young or not yet born in 2001 and the attacks have become less relevant. Also, the phenomena of a war weary country resulting from wars in Iraq and Afghanistan naturally leads to a reaction that rejects the idea that the world is a dangerous place

V. Oxford (✉)
Pacific Northwest National Laboratory, Richland, WA, USA
e-mail: Vayl.Oxford@pnnl.gov

S. Apikyan, D. Diamond (eds.), *Nuclear Threats and Security Challenges*,
NATO Science for Peace and Security Series B: Physics and Biophysics,
DOI 10.1007/978-94-017-9894-5_23

and we must be vigilant against future aggression. It has become very difficult to generate the passion and constancy of effort we felt after 9/11.

However, we cannot ignore the evidence in front of us that not only is the world a dangerous place but all of the signs point to a geopolitical environment that is becoming both more complex and more dangerous resulting in the world facing unparalleled challenges than any in our history. While our history shows that we have been up to the task of responding to threats, it also suggests that our reaction is almost always in response to an attack rather than aggressively anticipating and preventing the attack. Because of the catastrophic and unaffordable consequences of a nuclear attack, we cannot wait to respond but, instead we must heed the evidence and work diligently to prevent the attack.

There are four principal factors that point to an emerging threat environment unlike any we have faced before. This paper intends to highlight those four factors, their implication on national and global security and identify some steps that we need to take to be prepared. The four factors are:

1. The dissolution of nation-state control and the changing conditions in the Middle East and North Africa (MENA),
2. The consequences of troop withdrawal from Afghanistan and Iraq,
3. The impact of Iran's nuclear program and its future status as a nuclear weapons state and,
4. The implications of Russian aggression in Crimea and the Ukraine on global security.

In each case, these factors will challenge U.S. and other nations' application of their elements of national power (e.g. diplomacy, military options, intelligence collection), but this paper will not address those in depth.

23.2 The Dissolution of Nation-State Control and the Changing Conditions in the Middle East and North Africa (MENA)

The MENA region is transitioning to an environment where affiliations to ethnical and cultural values are more important that loyalty to a national government and its control. The unrest in Egypt, Syria, Libya, and Iraq is likely to be the beginning of a larger migration of affinities to cultural interests and should be expected to be followed by similar unrest in places like Jordan, Lebanon, Bahrain and elsewhere and will last for years if not decades. This sectarian violence represents a struggle for power that could have dramatic results including the potential for redrawing national borders based on historical and cultural lines. The dramatic success of ISIS in taking control of large amounts of territory consistent with historic lines, its ability to capture large amounts of military arms and financial assets and its stated desire to do grave harm against Western interests is a clear and present danger and cannot be ignored. Meanwhile, Al Qaeda is resurging and expectations are that they will be in a power struggle with ISIS to "out gun" them in executing terrorist attacks. Meanwhile terrorist groups and freedom fighters are flowing into these struggles to seek additional influence and power. The export of those freedom fighters to the U.S., Europe and other parts of the world represents and immediate and serious threat to global security. Moreover, the exposed Southern flank of Europe could result in increased ISIS and Al Qaeda attacks.

Meanwhile, the case in Syria is particularly important and could serve as a benchmark for what the future holds. In fact, DHS Secretary Johnson in a recent speech at the Wilson Center stated that Syria is a homeland security issue citing the potential for the freedom fighters to export their skills and ambitions to other parts of the globe including the U.S.[1] If the Assad regime survives, it is expected that internal brutality will continue, but more importantly, those that supported the regime including Iran/Hamas, Lebanon/Hezbollah, and Russia could emerge with increased status and influence. If the regime collapses, civil war will prevail for many years creating an environment for terrorist groups to expand their footprint while also being "lost in the noise" within the wider conflict and could lead to access to greater resources and expertise.

The erosion of nation-state control and authorities across the MENA region will challenge all elements of U.S. and partner national power resulting in the need to adapt diplomatic approaches to deal with an array of formal and informal government entities. Intelligence collection and analysis capability may be stretched beyond capacity demanding novel approaches to monitor a growing number of potential trouble spots. Military options may be limited based on denied access issues unless large ground forces become an option.

With respect to terrorist groups, this environment will offer opportunities for resurgence and to allow them to reinvent themselves to be less visible and to integrate into the new societal "norm" to gain access to resources.

23.3 The Consequence of Troop Withdrawal from Afghanistan and Iraq

Concurrent with the dynamic changing environment across MENA is the impact of troop withdrawal from Afghanistan and Iraq. This will have tangible and intangible impacts. In fact, there is evidence showing that troop withdrawal leads to reduced U.S. and coalition influence in the region. As an example, Iraq has quietly allowed Iranian support to the Assad regime by permitting use of its airspace.

Expected fallouts from the combined withdrawal include the loss of situational awareness across the region and the degradation of military response options due to reduced force strengths on the ground. Senior military have expressed concerns about losing situational awareness that in turn will limit U.S. capability to keep

[1] "A Conversation with Jeh Johnson", The Wilson Center, February 28, 2014.

terrorist groups off guard and to disrupt operations. Similarly, some former senior intelligence officials have cited the decline in intelligence service cooperation with the U.S. that is diminishing intelligence collection and analysis.

Predicated on these outcomes, we have seen that more freedom of action is available to terrorist groups to regroup, acquire resources and gain momentum in planning and executing attacks. More of this should be expected. In fact, a Washington Post article on December 29, 2013 entitled "Grim future seen for Afghanistan" stated that according to those knowledgeable of the new National Intelligence Estimate (NIE) on the Afghan war, it "predicts that the gains made by the United States and its allies during the past 3 years are likely to have been significantly eroded by 2017, even if Washington leaves behind a few thousand troops and continues bankrolling the impoverished nation … ". Further the Post reports that the NIE "predicts that the Taliban and other power brokers will become increasingly influential as the United States winds down its longest war in history…"[2] The recently agreed to Status of Forces Agreement that allows approximately 18,000 U.S. troops in Afghanistan could mitigate some of these concerns but overall effectiveness will rely heavily on cooperation with the Afghanistan government and the effectiveness of its own military.

23.4 The Impact of Iran's Nuclear Program and Its Future Status as a Nuclear Weapons State

Iran's nuclear weapons program poses a separate set of challenges to U.S. national security strategy and global security in general, and while negotiations continue to look for progress in reversing the program, Iran's past behavior suggests that little progress should be expected. Moreover, Iran appears to have studied the sanctions process in North Korea and has learned accordingly. Recently the Director of National Intelligence testified that Iran's breakout posture would allow it to produce a nuclear weapon in months not years should it make the decision to do so.[3] The critical question is whether Iran makes that decision or whether Iran will be content in the near-term to remain in a breakout mode. In either case, this poses serious challenges to the global strategic balance and represents challenges to international norms. The situation is further complicated by U.S. policies to reduce its stockpile and to seek a "global zero" trajectory for nuclear weapons worldwide. Meanwhile, our extended deterrence assurances are being met with increasing skepticism as other nations become nervous about U.S. commitments. These reactions could drive others to assess their national security and to reconsider their own nuclear ambitions. In fact a 2013 public poll in South Korea revealed that 70 % of South Koreans

[2]"Grim future seen for Afghanistan", The Washington Post, December 29, 2013.

[3]Statement of Hon. James R. Clapper, "Hearing to Receive Testimony on the Current and Future Threats to the National Security of the United States", April 18, 2013.

wanted the return of U.S. nuclear weapons to the Korean Peninsula or that South Korea should develop its own capabilities.

The combination of Iran's nuclear program and the U.S. strategic posture could lead to others like Saudi Arabia, Egypt, and Japan to consider their own security needs and to consider nuclear options. This would further erode U.S. credibility and could lead to a serious challenge to the future of the Nuclear Nonproliferation Treaty.

From a terrorism perspective, Iran would represent the first nuclear weapon capable state with a direct nexus to terrorism and thereby posing a serious asymmetric nuclear threat to the U.S. and others. This would be a game changer with respect to traditional U.S. nuclear deterrence and national security strategy.

23.5 The Implications of Russian Aggression in Crimea and Ukraine on Global Security

The Russian annexation of Crimea and its aggression against Ukraine poses some interesting questions for nuclear stability. While still early in the evolution of this issue and its impact on nuclear proliferation, some interesting questions arise. Would Russia have taken such bold moves if Ukraine had retained what amounted to the world's third largest nuclear stockpile after the collapse of the Soviet Union? What is the message to other nuclear weapons states and those that may have aspirations like Iran? What is the message to those countries that are currently on the fence about pursuing such a capability? Will this provoke a discussion within the U.S. to alter its current stockpile reduction strategy and global zero ambitions in order to bolster our extended deterrence credibility and in turn, act as a hedge against possible wide spread nuclear proliferation? These are serious questions that require serious consideration.

Similarly, the ability to continue nonproliferation efforts will become more difficult if not impossible. If that turns out to be the case, will the nuclear security community have confidence that all nuclear weapons and material have adequate security and that none will fall out of Russian control? If there is any uncertainty in the answer to that question, what steps should the U.S., other nations or NATO more specifically take as a hedge against non-state actors or terrorist group from gaining access to weapons or material?

23.6 Proposed Next Steps

Some critical steps that the U.S. should consider as a means to mitigate the possible consequences of the emerging threat and geopolitical environment are:

1. Develop a whole of government approach to combat the expanding adversary and threat base to include:

(a) Balance and integrate our foreign and domestic security agendas. In the post-9/11 environment and with the establishment of DHS, we have yet to develop integrated strategy, plans and budgets to address threats to the U.S. To date, the two security agendas remain uncoupled and are planned independent of each other,
(b) Develop joint interagency campaign plans, CONPLANS and operational plans to clearly describe departments and agencies roles and responsibilities and to respond to nuclear threats,
(c) Accordingly, convey to State and local authorities the roles they would play in a nuclear crisis,
(d) Examine and implement options to expand the bandwidth of intelligence collection and analysis capabilities to account for an expanding adversary base and concurrently, streamline information sharing across the Federal level and down to the State and local levels,
(e) Develop mechanisms to share situational awareness information across the Executive Branch, and
(f) Conduct senior-level exercises from the President on down to evaluate the effectiveness of the plans and to identify gaps in capability as well as issues with authorities.

2. Prepare for and enhance capability to defeat asymmetric threats to include better integration of OCONUS and CONUS assets, capabilities and information sources.
3. Take steps to better understand the cultural values, motives and intentions of a diversifying geopolitical environment in order to dissuade potential hostile intent and to better assess the threat space.
4. Rethink the national security policy structure to include consideration of re-establishing a Homeland Security Council structure to facilitate the planning, budgeting and integration of foreign and domestic security strategies and to achieve better integration across the array of domestic agencies.
5. Based on integrated foreign and domestic needs restore defense and homeland budgets accordingly.

With respect to international impacts of this new environment, consideration should be given to developing options and approaches to deal head-on with the "new norm" to include exploring:

1. A NATO/European "reset" with respect to Russia and establishing new security measures to protect against loss of control of Russian nuclear material or weapons.
2. New nonproliferation approaches to detect/interdict weapons or material
3. Regional containment strategies focused on key nations or regions of concern
4. Acceleration of efforts under the Global Initiative to Combat Nuclear Terrorism to enhance collective security of its member states to include nuclear forensics and attribution, nuclear detection deployment, exercise and training and information sharing

5. Means to extend nuclear deterrence assurances across the Middle East region and elsewhere as appropriate
6. Improved joint intelligence sharing focused on possible nuclear acquisition by non-state and terrorist groups
7. Incorporating NATO into U.S. CWMD campaign planning

23.7 Conclusion

In summary, the world is become both more complex and dangerous and the nuclear threat is likely to adapt to this new environment and now is not the time for complacency. Bold action is needed along with sustained national and international leadership to prevent a catastrophic nuclear attack anywhere in the world.

Chapter 24
Cross-Border Cooperation: A Key Towards Better Preparedness for Nuclear and Radiological Threats

Anna Nalbandyan

Abstract This article gives an overview of joint efforts done under a few NATO-supported projects which were implemented at different times in different regions: the South Caucasus, Central Asia, Belarus (the Chernobyl Exclusion Zone) and undertaken in the Euro-Arctic region, too, with the goal to improve and consolidate the cross-border cooperation in emergency preparedness and radiation protection, joint risk assessments and exercises as well as to enhance public awareness and knowledge in cooperation regions. The article also addresses some common challenges in cross-border cooperation pinpointing issues that need to be better harmonized for better international preparedness.

24.1 Introduction

For recent decades significant actions have been undertaken and many resources invested in many countries at a national level to improve emergency preparedness and response, radioactivity monitoring and risk assessments to more effectively protect environment and human health against nuclear and radiological threats.

However, a cross-border nature of any potential radioactive contamination and common challenges in border countries urge to consider radioactivity-related risks, monitoring approaches, environmental and human health protection strategies not only at a national, but also a regional and international level.

One of prerequisites for combating nuclear and radiological hazards is an efficient information exchange between countries. Besides, to increase understanding, trust and the potential to jointly mitigate challenges and associated risks as well as to assure quality and comparability of shared data, it is extremely important to build direct collaboration between key authorities and research organizations in emergency preparedness, security and protection around the world.

A. Nalbandyan (✉)
Norwegian Radiation Protection Authority, Tromsoe, Norway
e-mail: Anna.Nalbandyan@nrpa.no

S. Apikyan, D. Diamond (eds.), *Nuclear Threats and Security Challenges*,
NATO Science for Peace and Security Series B: Physics and Biophysics,
DOI 10.1007/978-94-017-9894-5_24

Another essential aspect to focus on is public awareness and knowledge essentially based on the principle of interaction with the society not only during and/or after accidents, but also in 'peaceful' times in order to share basic knowledge on radioactivity, radiation protection and associated risks. Timely provided information helps improve understanding and thus noticeably decrease the level of fear and panic among the general public and stakeholders.

24.2 Background

The idea of this article was inspired by several cases of efficient cross-border cooperation built in the frames of a few NATO-supported projects which were implemented in the South Caucasus, Central Asia (Tajikistan) and Belarus (the Chernobyl Exclusion Zone), and a EU-supported project which covered the Euro-Arctic region (Norway, Finland and Russia). Cooperation in each of the regions had its own history, aspirations, challenges, specificities and topics of interest. The author was directly involved in these projects and so would like to share the major achievements of cross-border cooperation in the studied regions, address main issues and joint solutions and underline some areas for further improvements.

24.2.1 *Region I: The South Caucasus*

The research goal was to monitor the basin of Rivers Kura-Araks – the largest water artery and the major environmental concern to the South Caucasian region (Fig. 24.1). The rivers are transboundary and thus strategically, politically and economically essential to all the three South Caucasian republics: Armenia, Georgia and Azerbaijan [2].

Fig. 24.1 A map of the South Caucasus river network (*left*) and Metsamor NPP (*right*)

This unique cooperative long-term research was implemented in the frames of a NATO Science for Peace (SfP)/OSCE project N977991 ‘South Caucasus River Monitoring’ (2002–2007), partner countries: Armenia, Azerbaijan, Georgia, USA, Belgium and Norway.

A rationale behind continuous monitoring and assessment of radioactivity of Rivers Kura-Araks basin waters and controlling possible radioactive pollution was a presence of an operating Metsamor NPP in Armenia (the only one in the region as of 2002), planned exploration of U deposits in the country, regionwide development of nuclear technologies, intentions to erect NPPs in neighboring countries.

The biggest challenge for cooperation in the South Caucasian region was unstable military and political situation between Armenia and Azerbaijan and lack of trust. Another challenge was lack of a water quality management system (in terms of specific pollutants such as heavy metals, POPs and radionuclides), modern instrumentation and analytical procedures as well as application of international QA/QC standards in the three South Caucasian republics.

24.2.2 *Region II: Central Asia (Tajikistan)*

The collaboration was built with an intention to assess environmental radioactivity status of former uranium mining and tailing sites in the north of Tajikistan (Fig. 24.2).

Uranium ore mining and processing in the former Soviet Republic of Tajikistan was initiated after the World War II. One of consequences of those activities have been huge amounts of generated uranium tailing materials and waste rock deposits, often found dumped close to inhabited areas. Key sources of concern have been technologically enhanced levels of naturally occurring radionuclides (TENORM).

In August 2008 a joint field mission was taken as a part of a ‘Joint project between Norway, Kazakhstan, Kyrgyzstan and Tajikistan: Environmental impact assessment of radionuclide contamination of selected uranium mining and tailing sites’ (coordinated by the Norwegian University of Life Sciences and funded by

Fig. 24.2 A map of the Central Asian region (*left*) and the Republic of Tajikistan with indication of the 2008 field mission sites (*right*)

the Norwegian Ministry of Foreign Affairs), and a NATO SfP project N981742 'RESCA – Radioactivity Environmental Security in Central Asia' (coordinated by the Josef Stefan Institute, Slovenia).

Given a lack of a proper waste management in most of former uranium mining and tailing sites in the study region, there exists a considerable risk of contamination spread far beyond the limits of presently known contaminated sites, making thus nuclear and radiological risks and potential transboundary contamination an international challenge [3].

24.2.3 Region III: Belarus (Chernobyl Exclusion Zone)

A study site for this cooperative research was the Belorussian sector of the Chernobyl Exclusion Zone (CEZ). The overall goal for international cooperation in this area (Fig. 24.3) was to obtain comprehensive information on radioactive contamination of the territory of the Polessie State Radiation-Ecological Reserve (PSRER) and to monitor transboundary transport of contamination to the territory of Belarus as a result of extreme natural phenomena, such as fires and flooding [1].

This joint research was done in the frames of a NATO SfP project N983057 'Polessie State Radiation' in partnership with Belarus, Ukraine and Norway.

Although numerous radioecological studies were conducted within the CEZ in the years following the Chernobyl accident of 1986, nonetheless there were significant uncertainties associated with mapping of radionuclide distribution and transboundary contamination [1].

Fig. 24.3 The Polessie State radiation ecological reserve (Brown et al. [1])

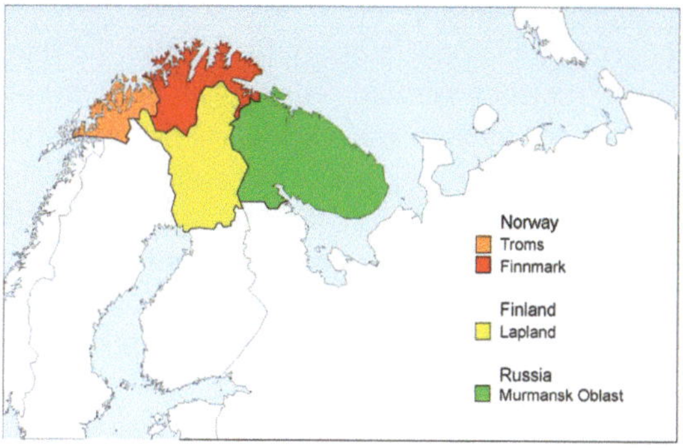

Fig. 24.4 A joint project site in the Euro-Arctic region (*left*) and sources of radionuclides in the northern marine environment (*right*, NRPA [7])

24.2.4 Region IV: Euro-Arctic

The Euro-Arctic region has become a region of international importance. Environmental radioactivity is of special concern to the Arctic region due to numerous existing and potential sources of radioactive pollution in immediate and adjacent areas (Fig. 24.4). Besides, sub-Arctic and Arctic food chains are short, this providing a potentially quick uptake route for contaminants to reach biota and the man. This fact, too, supports a necessity to continuously monitor environmental radioactivity particularly in areas which are actively used by local people for collection of natural food products ([4, 5, 9]).

The three partner countries: Norway, Finland and Russia have already developed their own requirements to emergency preparedness, radiation protection and environmental monitoring [8]. To jointly mitigate challenges in the Euro-Arctic region, it is essential that direct collaboration be built between key authorities, research organizations and stakeholders in the three partner countries and procedures and methodologies be harmonized to strengthen emergency preparedness capabilities in the High North. It is also important to get comparable data for environmental and other risk assessments for the entire Euro-Arctic region vs. restricted assessments for each country. So, we made a suggestion that a cross-border cooperation network be built and areas identified that would focus on specific issues and challenges, and later implemented it the frames of the CEEPRA KO130 project 'Collaboration Network on Environmental Radiation Protection and Research' under financial support of the EU Kolarctic ENPI CBC Programme and the Kolarctic Norge.

24.3 Methods

As pointed out earlier, each of the regions had specific challenges, needs and topics for international cooperation. Besides, the regulation, monitoring procedures and emergency preparedness approaches were also different in all partner countries.

A prerequisite for successful implementation of the four described projects was development of a specific cooperation plan and strategy for each of the four regions that would both meet the joint project goals and comply with demands of local authorities. An essential aspect of cooperation was identifying pathways for harmonization of approaches and a possibility to get credible and comparable data for combined environmental and other risk assessments. Also, cooperation included improvement of analytical procedures, measurement techniques and modeling capabilities in partner countries, joint field works, practical experience exchange and inter-comparisons (both in-situ and inter-laboratory), discussion of approaches for subsequent follow-ups.

Another aspect of implemented cooperation was interaction with local society, sharing information and knowledge on specific issues in a popular way.

24.4 Results and Discussion

24.4.1 Region I: South Caucasus

Due to the cross-border cooperation in the South Caucasus a joint assessment of the Rivers Kura and Araks basin radioactivity was performed and agreements achieved between the three South Caucasian partner countries. This first ever cooperative research of such a kind contributed to the security and safety of the region, increased trust between researchers from the partner countries and enhanced preparedness capabilities in the case of emergencies in the region. Major specific achievements were that:

- The environmental laboratories from Armenia, Azerbaijan and Georgia have got identical modern instrumentation for radioactivity measurements;
- To assure data credibility and the quality of the research, a Quality Assurance and Quality Control Project Plan and standard operational procedures based on ISO general requirements to sampling methods and a guide for water sample treatment and measurements were developed;
- Special protocols were developed for registration of ultimate outcomes on all stages of river water monitoring: from water sampling to final data analysis.
- A joint database was compiled which included data for all the three South Caucasian republics.

Through the entire project period, river water sampling was done synchronously in all the three South Caucasian republics. A joint project website was created where monthly data on radioactivity measurements from the three countries was shared.

24.4.2 Region II: Central Asia (Tajikistan)

The mission to northern Tajikistan allowed to jointly assess the current environmental radioactivity status of legacy uranium mining and tailing sites in Taboshar and

Digmai. Over time, such tailings undergo chemical reactions and therefore their mineralogy and pore water composition may potentially change. To understand the fate of radionuclides, continuous monitoring and measurements are required [3]. In Taboshar assessments covered the so-called Dead Lake – a former uranium extraction site with an artificial lake originated as a result of mining activities, and a uranium tailing pile emerged in the result of low grade uranium ore processing (U content below 0.03 %). In Digmai monitoring was performed for the uranium mill tailing dump – a 90 ha area with dumped radioactive waste generated as a result of uranium ore processing. The quality and compatibility of field measurement data were assured due to in-situ inter-comparison of measuring devices (a dose rate meters, radiometers and digital hand-held spectrometers). The major results obtained for Tajikistan include in particular:

- The revealing and mapping of areas having highest radioactivity levels, detection of radioactivity changes around the Dead Lake, Taboshar uranium tailing pile and Digmai repository site;
- γ-spectrometric determination of radionuclide composition of the collected environmental samples and current distribution and ratios of radionuclides;
- Preparation of a crosscutting joint report on the status of radionuclide contamination, impacts and risks of the former uranium legacy sites in Tajikistan.

It should be stressed that the Digmai area has been labeled a hazardous zone, nonetheless it remains open to the public; another area – the Dead Lake – is often used as a grazing site for local animals. The results obtained could be used by local authorities as a basis when evaluating and selecting risk mitigation measures and site rehabilitation techniques [3].

24.4.3 Region III: Belarus (Chernobyl Exclusion Zone)

Collaboration of scientists from Belarus, Ukraine and Norway has added considerably to our understanding of the distribution, behavior and fate of radionuclides in the Belarusian sector of the CEZ [1]. As a result of cooperation:

- Uniform procedures and methods for sampling and the determination of radioactive contamination were developed for the study area;
- Data assessment was done employing geo-statistical methods;
- The sampling and mapping data from Belarus and Ukraine were combined and results compared;
- A database was collated and digitized maps produced to show radioactive contamination of the PSRER territory;
- Models were validated for simulation of ^{137}Cs and ^{90}Sr migration in the terrestrial landscapes of the Exclusion Zone and its vicinity [1].

24.4.4 Region IV: Euro-Arctic

As a result of cooperation a CEEPRA network was established between relevant authorities and research organizations from Norway, Finland and Russia [8]. The collaboration areas were terrestrial and marine environment monitoring, atmosphere (hypothetical accidents and modeling), social impacts assessment and public awareness and knowledge. The established network will serve as a platform for further sustainable cooperation and exchange of information and expertise in the Euro-Arctic region. The main outputs of this cooperation:

- Improved emergency preparedness capabilities in the region: region-specific risk assessments were carried out through modeling and investigation of long-term effects of potential nuclear accidents in the Euro-Arctic region and possible impacts on region's indigenous population, environment, reindeer husbandry, the natural product sector, tourism and industries;
- Improved modeling: atmospheric and marine models were compared between countries and capabilities improved for better preparedness in the region;
- Joint environmental assessments: the current and long-term state of radioactive contamination in terrestrial and marine ecosystems in the Euro-Arctic region was assessed. A special emphasis was placed on natural food products widely used by population such as berries, mushrooms, fish and reindeer meat. For the first time joint maps were produced employing data from the three countries;
- Information to the public: open seminars for general public and target groups were arranged in Norway (Tromsø), Finland (Oulu) and Russia (Murmansk) in 2011–2013 and information shared on different radioactivity-related issues.

24.5 Common Challenges and Issues to Be Harmonized

As shown by a 12 year long cooperation in different geographical regions and in different areas (such as emergency preparedness, radiation protection, monitoring and risk assessments, public awareness and knowledge), there exists a strong need in consolidation of cross-border cooperation and harmonization of current approaches between different countries to be better prepared and capable to prevent, detect and adequately respond to a an emergency situation and a nuclear or radiological threat as well as to deal with the legacy sites.

Today, challenges are very different from those defined almost half a century ago, threats and prevention and decision making technologies, too, have been changing steadily. This dictates a necessity to develop new improved decision support tools, equipment and skills, new risk models and instructional procedures.

In the epoch we are living, international mass media have become widely accessible, so the local authorities must be able to operatively respond to any incident or accident and provide proper instructions and a prognosis of situation in a very short time in order to prevent spreading of contradictory information from other information sources.

For us as scientists and experts it is imperative and a common task to develop a common strategy on response to different emergency situations to efficiently manage challenges which may potentially emerge both in neighboring and remote countries. Here emphasized should be steps to be taken immediately after an accident as post-accident preparedness is very – not to say vitally – important! A specific challenge here is to bring clarity into the roles and responsibilities of responders on each level. On the other hand, it is important to emphasize that the post-accident issues are not primarily radiation issues! [6].

Effective cross-border cooperation helps to satisfy another need: giving and taking a prompt cross-border advice in case of any nuclear or radiological emergency. Besides, some countries are more advanced and have a better expertise in specific areas, so sharing experience and knowledge is exceedingly important, too.

Another question is whether we have a common understanding of risk. This could have different meaning for different target groups. For instance, scientists consider assessments of possibilities and consequences, politicians – emergency preparedness and costs, the general public – food contamination and public health, industry specialists – economic aspects.

The public perception of radiation-related risk is also very different and sensitive, public education and communication are equally important both in 'peaceful' times and on all stages of declared emergency situations as there exists a risk that public will not wait for instructions and will evacuate themselves in case of any accident. So, along with basic knowledge and understanding of radiation and related issues, public instruction should involve practical exercises in actions to be taken in case of potential emergencies. The public instruction should involve practical training for different emergency situations. And finally, 'to avoid the social breakdown and chaos it is important to build and maintain trust' (Rafael Vidal, Mayor of Asco, Spain).

A challenge many countries face today is lack of defined strategy and effective communication of radiation protection knowledge with public, lack of educational programs, popular lectures and brochures for different target groups. On the other hand, a huge amount of thematic instructional courses and e-material developed by different countries has been generated on the Internet, but the scattering of such information all over the immense net space strongly complicates its search. So, it is advisable that existing material be compiled into a single, unified source (a website) that would help overview all available educational possibilities and locations. Another suggestion is establishing regional or national Centers for Practical Radiological Culture that would contribute to improvement of public awareness and preparedness.

24.6 Conclusion

Bordering countries face common or almost similar challenges, so, cross-border cooperation assures wider possibilities to build effective systems for emergency preparedness and consolidate national, regional and international capabilities in detecting and preventing nuclear and radiological treats, defining strategies to adequately respond to and mitigate consequences in any emergency situation or legacy site management.

Joint cross-border exercises considerably improve practical abilities and skills in partner countries through exchange of expertise and knowledge.

Effective collaboration of countries – both neighboring and remote – helps harmonize procedures, develop common strategies, get comparable data, build trust and get a quick feedback if necessary.

Effective public communication helps refine the public vision, reduce the level of panic in population in case of emergencies and identify credible sources of relevant information both for general use and self-instruction.

Acknowledgments I wish to express my sincere thanks to the colleagues whom I cooperated with in the four projects for their dedicated efforts that made the cooperation successful. We gratefully acknowledge support from the NATO and OSCE, EU's Kolarctic ENPI programme and Kolarctic Norge, Norwegian Ministry of Foreign Affairs and coordinating institutions from Armenia, Georgia, Azerbaijan, Tajikistan, Belarus, Ukraine, USA, Belgium, Slovenia, Finland, Norway and Russia.

References

1. Brown JE, Bondar Yu, Kashparov V, Nalbandyan A, Navumav A, Skipperud L, Sokolik G, Strand P, Zhukova O (2011) Radioactive contamination in the Belarusian sector of the chernobyl exclusion zone. International conference on radioecology & environmental radioactivity. 19–24 June 2011, Hamilton. Paper N 948, 6 pp
2. Nalbandyan A, Saghatelyan A (2008) Radioecological monitoring to the South Caucasus' rivers as a tool for early warning system. In: Apikyan S et al (eds) Prevention, detection and response to nuclear or radiological threats. NATO science for peace and security series, vol 2. Springer, Dordrecht, pp 183–190
3. Nalbandyan A, Hosseini A Environmental radioactivity assessment at nuclear legacy sites in the Republic of Tajikistan. Abstracts, the third European IRPA congress 2010, 14–18 June, Helsinki – Publ. NSFS-Nordic Society for Radiation Protection, p 246
4. Nalbandyan A, Gwynn JP, Eikelmann IM, Møller B, Nilssen J, Rudolfsen G (2012) Assessment of environmental radioactivity in natural food products from northern Norway. Abstract (A2357886) and Poster (P11.41), the 13th International Radiation Protection Association (IRPA) Congress 'Living with Radiation – Engaging with Society', 14–18 May 2012, Glasgow, Scotland (electronic)
5. Nalbandyan A, Gwynn JP, Møller B, Nilssen J, Leppänen A-P, Rasilainen T, Kasatkina N, Usiagina I (2014) Quality assurance in cross-border monitoring of environmental radioactivity in the Euro-Arctic region. Abstract and poster, session: PS7-2 – the fourth European IRPA congress, radiation protection culture – a global challenge, 23–27 June 2014 (A), Geneva. http://www.irpa2014europe.com/abstract/

6. Nalbandyan A, Gwynn JP, Møller B, Leppänen A-P, Rasilainen T, Kasatkina N, Usiagina I (2014) Cross-border assessment of environmental radioactivity in the Euro-Arctic region. Abstracts, ICRER 2014 – third international conference on radioecology and environmental radioactivity, 7–12 September 2014 (B), Barcelona. Oral and poster presentation N 296. https://intranet.pacifico-eetings.com/amsysweb/publicacionOnline
7. NERIS-TP (European platform on preparedness for nuclear and radiological emergency response and recovery) dissemination workshop, 22–24 January 2014, Oslo
8. NRPA (2007) Radioactivity in the marine environment 2005. Results from the Norwegian National Monitoring Programme (RAME). StrålevernRapport 2007:10. Østerås-NRPA,42 pp
9. Rasilainen, T (ed) CEEPRA – collaboration network on EuroArctic environmental radiation protection and research. Final report. Säteilyturvakeskus, Kopijyvä Oy, Jyväskylä 2014, 44 pp. ISBN 978-952-309-002-6

Chapter 25
Challenges in National Nuclear Threat Reduction Field-Regional Puzzle

Lia Chelidze and Giorgi Nabakhtiani

Abstract Georgia had received difficult heritage form Soviet related to nuclear security situation. Great problem connected to so-called orphan radioactive sources. The most significant orphan radioactive sources found in Georgia are $^{90}Sr/^{90}Y$ used for special thermogenerators of electricity. Initial activity of each one was 1295 TBq. The country already took some steps to combat nuclear and radiation threat. Due to the geostrategic location of Georgia and whole Caucasus region formation of specific approaches towards radiological and nuclear threats is essential. Georgia is continuing its cooperation with the IAEA. As a result of the close cooperation the national Integrated Nuclear Security Support Plan (INSSP) was developed.

25.1 Introduction

New Great Silk Road crosses Georgia, connecting Europe and Asia. Along this route oil and gas pipelines, transport and telecommunication lines will be laid. Unfortunately, besides economical communication, the route can also be used for illegal transit of nuclear materials. There is a special concern regarding uncontrolled territories in conflict zones. Safety of this route is a precondition of economical development and political stability of the whole Caucasian region.

25.2 Radiological Emergencies

The use of nuclear technologies opened a new era for mankind. It is very important to use these technologies safely; therefore radiation protection standards shall be established to govern application of ionizing radiation to provide necessary safety and security level. Safety culture that is shared by all is one of the indicator of well

L. Chelidze (✉) • G. Nabakhtiani
Department for Nuclear and Radiation Safety, Ministry of Environment and Natural Resources Protection, Tbilisi, Republic of Georgia
e-mail: lia.chelidze@gmail.com

S. Apikyan, D. Diamond (eds.), *Nuclear Threats and Security Challenges*, NATO Science for Peace and Security Series B: Physics and Biophysics, DOI 10.1007/978-94-017-9894-5_25

done preventive measures on country level. Georgia's nuclear programme is limited to typical uses in the medicine, agriculture, industry and research. Nevertheless, a potentially hazardous radiological situation has developed in Georgia in connection with so-called orphan radiation sources since 1997. Uncontrolled radioactive sources were highly radioactive and in most cases were located near the populated areas, they were a serious hazard for the population. There were found more than 300 sources being out of regulatory control in Georgia. The most recognized "orphan" radioactive sources refer to s.c. thermoelectric generators (RTG) based on ^{90}Sr. Initial activity each of them was 35,000 Ci. In 1990s of last century Georgia faced complicated problems in nuclear and radiation safety field, namely: weakness of infrastructure for state management, incomplete inventory of radiation sources, orphan sources all over the post soviet military bases and as a result several radiological emergencies, overexposed victims and high risk of attracting international terrorists to use the country as a rout for illegal movement of radiation sources or nuclear materials. Short description of the accidents by consequence; in 1993 in the town of Zestafoni (West Georgia) 41 undocumented ^{137}Cs sources were found. The sources used for calibration of radiometric instruments for military purposes were found in the manufacturer's shielding. A group of specialists from the Institute of Physics of the Georgian Academy of Sciences examined, identified and transferred the sources to a secure storage location. Kutaisi, West Georgia, 1996 – plundering and opening of the container, which held a powerful source of radioactive cobalt. Consequence – Injury to four persons, one of whom died due to the acute radiation 3 weeks after the incident. The high power "orphan" radioactive sources caused to so called Lilo, Khaishi and Tsalenjkha Radiological emergencies developed in Georgia [1, 2]. It is notable that in 1990s emergency situations were considered from the point of view violation of radiation safety norms resulted in the damage of health. After September 11 attacks the vector is coursed to terroristic threats sections.

25.2.1 Lilo

The Lilo Training Centre is located at a remote about 25 km east of Tbilisi. The centre covers approximately 150,000 m^2. In the past the centre was used by Soviet troops for civil defense training. The sources may have been used for calibration of survey equipment and for training in radiological monitoring to be carried out in the event of a nuclear accident or nuclear war. The site was transferred to Georgian Army in 1992. Chronology of the accident April–August 1997 [1, 2]: Soldiers from the Lilo Center developed skin burns on several parts of their bodies and army medical doctors initially are unable to determine the cause of lesions. There is no information on the time when the exposure began. From the Chemical, Radiological and Biological Protection Division of Soviet Army on 26 August 1997: A radiation hot spot was discovered at the Lilo Center near the underground shelter.

Table 25.1 Summary of the sources (Cs-137) found at the lilo training centre (GBq for ^{137}Cs)

Source number	Date found	Estimated activity (GBq)
1	1997-09-13	164
2	1997-09-13	126
3	1997-09-13	0.37
4	1997-09-14	0.88
5	1997-09-19	0.02
6	1997-09-19	0.02
7	1997-09-19	0.02
8	1997-09-19	0.63
9	1997-09-19	0.01
10	1997-09-20	0.02 -

The dose rate was about 45 mGy/h. 10 September 1997: The Center of Applied Research of the Institute of Physics assessed the radiological situation at the site. On 11 September 1997 was started monitoring of area in difficult preconditions – No information was available on the type of radionuclide, location etc. Totally 250,000 m^2 were monitored. An investigation had revealed that several ^{137}Cs, 60Co radiation sources, about 200 units of night shooting guides containing ^{226}Ra and some beta emitters had been found and that in some places high dose rates had been detected. The Government of Georgia requested the IAEA to send an emergency team to evaluate the radiological situation at the Lilo Training Center. Results are as detailed in Table 25.1.

The IAEA expert mission took place from 11 to 14 October 1997 to verify the dose rates at the Lilo Training Centre. The conclusion is a BSS is living document when the implementation of standards is prime responsibility of "legal" person: the first of all the previous owners of Lilo Training Centre and acting management. Unfortunately the comprehensive inventory of radioactive sources did not been transferred to national authorities by previous owners.

25.2.2 The Radiological Accident in Khaishi Region

After the accident in Lilo the Georgian Government, following the recommendations of the IAEA, initiated a programme to survey other similar sites with the aim of finding any other abandoned radioactive sources. In late July 1998, during implementation of this programme, more abandoned sources were found in the village of Matkhoji, in the region of Kutaisi, located 300 km west of Tbilisi. On 10 October 1998, two unidentified radiation sources of unknown origin with extremely high activities were discovered near the mountain village Idliani (Khaishi Region, Western Georgia). One of these sources was found without any shielding giving rise to a radiation dose rate of 50 mGy/h at a distance of 2 m from the source.

On October 14, 1998, after analysis of published data about radiation sources produced in the former Soviet Union, it was assumed that the sources found near the village of Khaishi could have been part of the so-called radionuclide thermo-electric generator group. In such generators different types of radioisotopes were used, whole activity can vary from 740 GBq- to 5,550 PBq (20,000–150,000 Ci) or more. In particular, as is pointed out in the 1979 edition of "Problems of Atomic Science and Technology" by A. M. Petrosiants, it demonstrates that various types of generators were designed on the basis of different radioisotopes: ^{144}Ce (with activity of 20,000 Ci); ^{90}Sr-(with activity of up to 100,000 Ci) and ^{137}Cs with activity from 50,000 to 150,000 Ci). These generators were used as sources of electrical power for radiometric devices and for navigational systems. At the request of the Georgian authorities the International Atomic Energy Agency sent a mission to assess the radiological emergency in Khaishi, to provide advice to the counterpart and to participate in the recovery operations for the reported radioactive sources. The design of the protective container for the transporting of sources was undertaken by specialists from the Institute of Physics of the Georgian Academy of Sciences. The container, containing both sources, had to comply with category two of the transport of radioactive materials regulations (the dose rate on the outer surface of the container could not exceed 0.5 mSv/h). In RTGs the strontium sources serve as a heat sources. The products of radioactive decay (Beta-particles) are captured inside the sources. They emit only "bremstrahlung". So it was suitable to use lead as a shielding material. According to the calculations, the thickness of the protective lead layer had to be 18 cm and the total weight was estimated to be about 1,300 kg.

25.2.3 The Radiological Accident in the Tsalenjikha Region

The next (third) pair of strontium sources was found on 23 December 2001. Their existence was discovered after three inhabitants (men) from the village of Zemo Lia, Tsalenjikha region were brought to the Institute of Hematology and Transfisiology in Tbilisi [3]. Two of them had heavy radiation burns on their backs and the third one to his hands. According to the men, they were collecting wood in the forest when they came across two unknown metal objects which were hot and for some period of time they used them as a heat source to keep themselves warm.

25.2.3.1 Radiological Assessment

The first attempt to determine the exact location of these sources and to examine their condition failed because of the almost impassable road and bad weather. The second attempt, made on 29 December 2001, was more successful. A team, consisting of specialists from the Nuclear and Radiation Safety Service, the Central

Department of Emergency Situations and the Institute of Physics, reached the place. They established the exact location of the sources, the sources' condition and made all necessary measurements, took photos and made a video film. The sources were located in a barren unpopulated region. The road, better defined as a dirt track, leading to the sources' location was practically impassable. This dirt track could only be used by woodcutters having high-powered three-axle cross-country vehicles with very experienced drivers who knew exactly all the difficult sections and specifics of the track. The last 400 m to the sources' location was absolutely impassable as it was blocked with rocks from a landslide. These sources were located 28 km along this dirt track from the village of Zemo Lia, the village of the three men who found the sources. This dirt track is primarily used by local inhabitants and woodcutters. Photskhoetseri, however, is the nearest village and residential area next to the sources' location. Its location from the sources is 18 km along this dirt track road but only 4–5 km as the crow flies. Photskhoetseri is, on the other hand, separated from the sources' location by a mountain ridge, the high-altitude dam and a reservoir. It is worth noting, that the sources were deposited just off the dirt track, in a hollow and were isolated from this track by a heap of rocks and earth. Due to these factors the radiation on the dirt track side was partially shielded and the radiation dose rate on this track, even in close proximity to the sources, was not very high. Close to the sources, on the other side of the shielded dirt track, at a distance of about 5 m from the sources, the dose rate was 1,3 mSv/h, while for unshielded sources this value would be expected to be in the region of 80 mSv/h. This fact was very important, as it allowed the recovery team additional time to carry out their preparatory work (to repair the road, to park the vehicle loaded with the container and to arrange for recovery devices to be placed in convenient locations) under conditions of relatively low radiation doses for the personnel. A detailed plan of the activities of the recovery group determined the time intervals during which the operators were allowed to be at different distances from the sources.

25.2.3.2 Source Recovery and Temporary Storage

Recovery of the sources was carried out on February 2–3, 2002. By this time the whole preparatory work was completed fully. A series of training sessions were carried out and a group of operators-executors was chosen. The operators were chosen from among the personnel of the Central Department of Emergency Situations of Georgia. The works on recovery of the sources were carried out under bad road and climatic conditions. It took 3,5 h to pass 18 km from the village of Photskhoetseri to the sources' location. Most of the road was covered with new-fallen snow and it was only possible to travel along it by means of a towing tractor provided by the local authorities.

Fortunately, in all the described cases of the removal of high radioactive strontium sources their locations were isolated from the roads (approaches) with

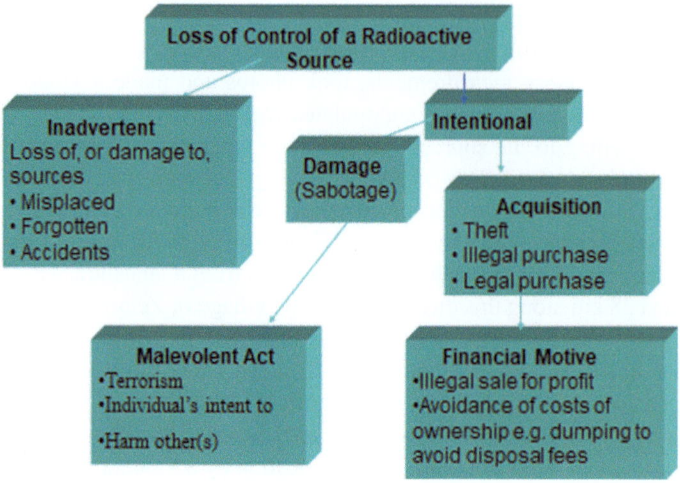

Fig. 25.1 Causes loss of control of a radioactive source

a heaps of earth or stones which served as shielding. That was very convenient for performing preparatory works (parking the vehicle with the container, arranging the auxiliary instruments according to the plan etc.). However, after the removal the sources from their initial locations they were placed in open areas, loosing their protective shielding and consequently exposing the operating personnel to increased dose rates.

In the end six of the strontium sources were found but there may be eight of them, or perhaps more. These small generators were used during the soviet era to power communication towers. Each contains a similar amount of the radioactive element strontium-90.

Following the notification of the first incident in Georgia, the Department of Technical Co-operation (TC) agreed to fully co-operate with the Department of Nuclear Safety (NS) for all organizational issues related to the dispatch of experts, staff and any relevant equipment. Some of the activities reported above were, therefore, implemented under existing TC mechanisms and projects in Georgia. The recovery of the highly active strontium sources was carried out with very complicated radiological, climatic and geographical conditions. Nevertheless the radiation hazard works carried out their tasks exactly according to the pre-defined plan and in less time than was expected and was done in the preparatory training. The doses received by the personnel were not so high and they were less then planned and expected. This was achieved by the correct planning of the operations, estimation of expected doses and good training in an atmosphere of co-operation. It should be mentioned that the quality of the operations increased from case to case. That demonstrates that lessons were learnt and taken into account (Fig. 25.1).

25.3 Different Scenarios

The experience of searching and recovery operation for orphan sources shows the possibility of development of different scenarios.

Simultaneously with strengthening of regulatory control, one of the effective methods to prevent any accident related to orphan radioactive sources is conducting of searching operations. There three main possibilities to conduct searching for orphan radioactive sources: Airborne Survey, Car Survey and Pedestrian survey. Airborne the most effective and expensive at the same time. This type searching was carried out in Georgia within the scope of IAEA TC project GEO/9/006 "Assistance for safe disposal of strontium-90 the thermogenerators" when 56 h of airborne gamma survey of a large territory of the western part of Georgia and around Tbilisi was carried out at 2000. Airborne survey is not effective for mountain relief. So, taking into account high price for this activity, car and pedestrian searching also were conducted at 2002, 2003 and 2005. All these activity were actively supported by the Agency in close collaboration with USA, France, Indian and Turkish experts.

We experienced two types of recovery operations: recovery during searching operation and large scale recovery. Large scale recovery operations usually contain three phases: Assessment, Identifying options for solution and Implementation.

Taking into account the geopolitical location of Georgia it is important to strengthen the physical protection infrastructure in country, which has territorial problems. Ever since the "dirty bomb" scenario became potential instrument of terrorists, this again points to necessity of strengthening physical control of radioactive materials. We believed in collective security as a guarantor of national safety. Political factors in the region which is located on the crossroads from north to south and from east to west, strategically linking together Asia and Europe may be promoting or inhibiting international cooperation in developing regional partnership capacity against the threat of nuclear terrorism.

Step by step Georgia established a regulatory body, worked out national legislation, developed licensing and inspection activities, installed radiation portals at the perimeter of the country. Several successful cases of prevention of the illicit trafficking of nuclear materials through borders of Georgia identified country's ability to contribute its shear to the Global Nuclear security.

Illustration for efficiency of established system: "Red Bridge" check point (east Georgia), when into scrap metal were fixed two well logging Pu-Be sources (Fig. 25.2).

Operations curried out against illicit trafficking of radioactive materials in Georgia for 2006–2013: in total 20 cases (2006-3, 2207-3, 2008-2,2009-2, 2011-2, 2012-6, 2013-2).

Fig. 25.2 Plutonium/beryllium source found in the scrap metal shipment at the "red bridge" on the Georgia-Azerbaijan border

25.4 Conclusions

There are some real general hazards connected with abandoned radiation sources, and illicit trafficking that became very significant due to geopolitical position of Georgia. Nuclear security geography ultimately defines a nation's interest and its ability to control events and activities. As an effective tool for in-state cooperation and capacity building as well as effective cooperation with partner states and international organizations, Georgia's Integrated Nuclear Security Support Plan (INSSP) is considered as a road map to achieve comprehensive approach to nuclear security regime best architecture. Georgia becoming a regional hub for nuclear security think-tank activities and cooperation by establishment of Regional Secretariat within the framework of the EU Chemical Biological, Radiological and Nuclear (CBRN) Risk Mitigation Centers of Excellences (CoE) Initiative for South East Europe, South Caucasus, Moldova and Ukraine. As a starting point, Georgia hosted "2014 CBRN Science and Consequence Management Congress" in June 2014.

The corresponding national regulatory infrastructure has been created for controlling the powerful radioactive sources to protect them against terrorism and theft on the territory of Georgia. Georgia is part of IAEA conventions. National integrated Nuclear Security Support Plan (INSSP) is developed for Georgia.

References

1. The radiological accident in Lilo//International atomic Energy Agency (IAEA), Vienna, 2000
2. Abramidze S, Katamadze N, Lomtadze Z, Cruz Sudrez R, Bilbao Alfonso AV (1998) The radiological accident in Tbilisi. Safety of radiation sources and security of radioactive materials contributed papers IAEA, Vienna, IAEA-TECDOC-1045. IAEA, Vienna, pp 131–133. ISSN 1011-4289
3. The radiological accident in Lia, Georgia//IAEA, Vienna, 2013

MIX
Papier aus verantwortungsvollen Quellen
Paper from responsible sources
FSC® C105338

If you have any concerns about our products,
you can contact us on
ProductSafety@springernature.com

In case Publisher is established outside the EU,
the EU authorized representative is:
Springer Nature Customer Service Center GmbH
Europaplatz 3, 69115 Heidelberg, Germany

Printed by Libri Plureos GmbH
in Hamburg, Germany